ASTRONOMY
from
A to Z

ASTRONOMY
from
A to Z

A Dictionary
of Celestial Objects and Ideas

Charles A. Schweighauser

Illinois Issues
Sangamon State University
Springfield, Illinois

Printed in the United States of America

Library of Congress Cataloging-in-Publication Data

Schweighauser, Charles A.
    Astronomy from A to Z : a dictionary of celestial objects and ideas / Charles A. Schweighauser.
        p.   cm.
    ISBN 0-9620873-5-1 : $14.95
    1. Astronomy–Dictionaries.   2. Astronomy–Terminology.
I. Title.
QB 14. S38   1991
520 ' .3–dc20
                                                                  91-7124
                                                                  CIP

Cover design by Larz Gaydos. Illustrations by Frank Thomalla, Judith Farris, Tom Gunter, Larz Gaydos, and Drena Stickel.

Cover front photo of Saturn by Voyager 1, 1980.   NASA Photograph.

Cover back photos:

Viking 2 photo of Mars (August 5, 1976) from a distance of 260,355 miles.   NASA.

On the left, Uranus's south polar region as viewed by Voyager 2 on January 17, 1986; on the right, false colors provide details of the same region of the planet.   NASA.

The giant planet Jupiter and two of its moons, Io (left) and Europa. The Great Red Spot is visible beneath Io. Voyager 1 photo on February 21, 1979.   NASA.

Illinois Issues
Sangamon State University
Springfield, Illinois 62794-9243

To my parents, Julia Schweighauser and the late Arthur Schweighauser, who provided me with the inspiration and the opportunity to study the Universe. Thank you.

Contents

ASTRONOMY
from
A to Z

Introduction

The nucleus of this book was a series of newspaper articles on a variety of astronomical topics published in several hundred weekly newspapers throughout Illinois. The articles were then compiled in a small volume that was made available to those persons who attended star parties at the Sangamon State University Observatory.

In the dozen years since the printing of the series of articles, astronomers have enlarged our knowledge of the Universe many times, and the series has grown into this book of descriptions of astronomical objects, ideas, and astronomers who have had major insights into the nature of our Universe. Also in recent years the Voyager spacecraft visited Jupiter, Saturn, Uranus, and Neptune, and revealed startling new information about these giant planets, their moons, and their rings. The Infrared Astronomical Satellite found more than two hundred thousand sources of infrared energy in the sky in less than a year. The Very Large Array radio telescope has mapped the sky in the radio part of the spectrum to unprecedented accuracy. New electronic detectors on large telescopes allow us to penetrate farther than ever into the Universe.

We have studied the Universe in visible light for eons—ever since human beings looked at the sky and wondered about what we were seeing. The other regions of the spectrum—radio, microwave, infrared, ultraviolet, x-ray, and gamma ray—have had to wait until appropriate technology was developed to explore the Universe in these wavelengths. Radio telescopes have now been developed that can monitor radio energy from the most distant galaxies in the Universe that we have been able to find. Earth-orbiting satellites carry extremely sensitive instruments above our atmosphere to study infrared and ultraviolet light from a variety of astronomical objects. X-ray and gamma ray telescopes also circle our planet, probing distant stars and galaxies for information about some of the most unusual and violent objects in the Universe.

Modern astronomy continues to reveal a Universe that was unimaginable to human beings only a generation ago. What we have learned in the past few years, however, is overwhelmed by the promise of what new instrumentation will tell us by the turn of the millennium in 10 years. The Hubble Space Telescope; new and extraordinarily sensitive x-ray, gamma ray, ultraviolet, and infrared telescopes; and radio telescopes that will literally span continents will provide more data and insights into the Universe in the next decade than we have learned in all the previous history of scientific exploration of the cosmos. It is, indeed, the most exciting time to be an astronomer and to have an interest in the astrophysical Universe.

There is no attempt here to cover all topics in astronomy. As the result of many years of teaching astronomy to college students, giving popular lectures on a wide range of astronomical topics to audiences of all ages, and answering questions from the many people who have participated in the star parties at the Sangamon State University Observatory, the astronomical topics that follow have been found to be the most interesting to people who have an enthusiasm for and a curiosity about the stars, planets, Solar System, galaxies, and other objects in the sky. If it inspires its readers to a deeper and richer understanding of the Universe of which we are a part, then it will have served its purpose.

"The scientist does not study nature because it is useful to do so. He studies it because he takes pleasure in it; and he takes pleasure in it because it is beautiful."   Henri Poincaré

A note about units

The metric system has been used throughout in order to conform with modern scientific practice.

Length

1 centimeter (cm) = 0.39 inch	1 inch = 2.54 cm
1 meter (m) = 39.37 inches	
1 kilometer (km) = 0.62 mile	1 mile = 1.61 km

Weight

1 gram   = 0.035 ounce
1 kilogram = 2.2 pounds

Temperature

Kelvin   = Celsius (Centigrade) + 273
Fahrenheit = 9/5 (Celsius) + 32

For clarity, length and temperature are occasionally expressed in both metric and English units.

Powers of ten

10^1 = 10 (ten)
10^2 = 100 (one hundred)
10^3 = 1,000 (one thousand)
10^4 = 10,000 (ten thousand)
10^5 = 100,000 (one hundred thousand)
10^6 = 1,000,000 (one million)
10^7 = 10,000,000 (ten million)
10^8 = 100,000,000 (one hundred million)
10^9 = 1,000,000,000 (one billion)
10^{10} = 10,000,000,000 (ten billion)
10^{11} = 100,000,000,000 (one hundred billion)
10^{12} = 1,000,000,000,000 (one trillion)
...and so on

10^{-1} = one-tenth (0.1)
10^{-2} = one-hundredth (0.01)
10^{-3} = one-thousandth (0.001)
10^{-4} = one-ten thousandth (0.0001)
10^{-5} = one-hundred thousandth (0.00001)
10^{-6} = one-millionth (0.000001)
10^{-7} = one-ten millionth (0.0000001)
...and so on

Exponential notation is often used to express a large or a small number. Hence, the speed of light is 3×10^5 kilometers per second (three hundred thousand kilometers per second), and a wavelength of blue light is 4×10^{-5} centimeters (four one hundred-thousandths of a centimeter).

Acknowledgements

I want to thank all those who worked on the book: Larz Gaydos, Tom Gunter, and Drena Stickel for the art work; Nancy Rachelle for word processing; and in the Institute for Public Affairs: Brenda Suhling, for word processing and trouble shooting; Janet Mathison, for copyediting; and Marilyn Huff, for editing and oversight.

Of course, I must also thank the hundreds of astronomers who have explained the astronomical universe in previous books and articles, and from whom I have borrowed graceful and succinct ways of expressing our science.

The Andromeda Galaxy

The Andromeda Galaxy is the farthest object that the human eye can see in the Universe without a telescope. Located at a distance of more than two million light-years from the Milky Way, it is barely visible in the constellation of Andromeda in the fall sky. (One light-year equals about 9.5×10^{12} kilometers or six million, million miles.) It appears as a fuzzy patch of light that can be seen only on a clear night without distracting urban lights or the moon.

It is a spiral galaxy similar in size and shape to our own Milky Way Galaxy, and is thought to be moving toward us between 10 and 267 kilometers per second. We do not have to fear a collision, however, because even at the higher rate it would take billions of years for the Andromeda Galaxy to reach us.

The giant Andromeda Galaxy, at a distance of 2.2. million light-years, is the closest spiral galaxy to our own Milky Way.

Palomar Observatory Photograph

The mass of the Andromeda Galaxy is on the order of three hundred billion Suns, similar to the mass of the Milky Way. There is a central bulge of old stars surrounded by the arms of the galaxy: a disk of younger stars, gas, and dust.

The Milky Way and Andromeda galaxies dominate the Local Group, a small cluster of about 20 galaxies that are bound together gravitationally. Most of the galaxies in the Local Group are small dwarf galaxies containing only a few billion stars each. The Local Group has a diameter of three million to four million light-years.

Asteroids

Also known as minor planets, these irregularly shaped chunks of rock and metal orbit the Sun in the same counterclockwise direction (as seen from north of the Solar System) that all the planets do. Reliable orbits for more than 3 thousand asteroids have been determined, and partial orbits for another 6 thousand asteroids have been calculated. It is estimated that an additional 20 thousand asteroids have been observed.

The first asteroid to be discovered, Ceres, was found in 1801; it is also the largest and therefore brightest asteroid with a diameter of about one thousand kilometers (six hundred miles). It takes more than $4\frac{1}{2}$ years to orbit once around the Sun.

Most known asteroids orbit the Sun in the asteroid belt, a wide zone from 2 to $3\frac{1}{4}$ astronomical units (three hundred million to five hundred million kilometers) from the Sun between the orbits of Mars and Jupiter.

There are two small groups of asteroids that are not in this belt. The Trojan Asteroids are found in two clumps in the same orbit as Jupiter, held there by the planet's large gravitational field. The Apollo Asteroids, about 30 of which are known, cross the Earth's orbit. Occasionally they pass close to the Earth. One recent near miss occurred in 1937 when Hermes went by us at a distance of 900 thousand kilometers (540 thousand miles), just about twice the distance of the moon from the Earth. In 1968, Icarus passed within six million miles of the Earth.

Were an asteroid the size and mass of Ceres, Hermes, or Icarus to hit the Earth, major disaster would result. It has been theorized that the apparent sudden disappearance of the dinosaurs 65 million years ago was the result of a collision of an asteroid or comet with the Earth.

The total mass of all known asteroids is estimated to be less than the mass of our moon. Asteroids are made of rock, mostly carbon and silicon compounds, and metal, primarily iron and nickel. Only five percent of the asteroids are metallic. A few asteroids appear to be binary, that is, two asteroid fragments revolving around a common center of mass. It is thought that many of the meteors we see in the nighttime sky and meteorites we find on the Earth are fragments of asteroids.

We know that asteroids in the asteroid belt between Mars and Jupiter did not originate from a planet that disintegrated, as was once thought, simply because there is not enough total material to make a planet. Asteroids are probably debris left from the origin of the Solar System 4.6 billion years ago. Jupiter's large gravitational field keeps these smaller bodies from accreting into one or more larger objects.

Astronomy

Imagine a primitive cave dweller hurling a crude spear or a rock at the moon. He throws his weapon not out of anger or fear, but out of a curiosity to see if he can hit it. To his surprise, and perhaps dismay, he cannot reach the moon or any of the other objects in the sky.

Ever since human beings developed the capacity for wonder, for having thoughts that went beyond the daily necessities of food and shelter, they have speculated about those shimmering points and globes of light that we now call the Sun, moon, and stars. The science of astronomy was born.

Ancient bones have been found that have curious markings, cut into them by our hunting and gathering ancestors, that apparently track the changing phases of the moon. These artifacts are at least 15 thousand years old, giving clear evidence that observations and records of celestial events make astronomy the oldest recorded science.

Babylonians and other peoples, living in the valleys of the Tigris and Euphrates rivers as long ago as 4000 B.C., were the first systematic recorders and interpreters of the sky. They identified many of the star groups, called constellations, that we still recognize today. They invented the zodiac—a band of 12 constellations against which the Sun, moon, and planets seem to move—as an aid in describing the positions of these objects.

Our rock-throwing ancestor who could not hit the moon had the same problem that astronomers have in the twentieth century: we never come into contact with the objects we are interested in studying. With the exception of meteorites that land on the Earth and the recovery of a few pounds of moon rocks, we will never study astronomical objects, such as a star or a galaxy, in a laboratory. Almost everything we know about the Universe we know due to our study of the light that the stars and planets send us.

Because these objects cannot be handled, taken apart, and rearranged, astronomy may be thought of as a pure science. Other sciences lead to applications—such as physics, which leads to atomic power—because the objects scientists are interested in studying can be manipulated. In the same way, geology leads to the use of mineral resources, chemistry to synthetic materials, and biology to hybrid corn. Except for passive uses such as navigation, astronomy has no practical applications.

Therefore, we study astronomy only for pure knowledge, to gain insight into and understanding of the organization of matter, energy, and the Universe. Astronomy leads us to seek answers to the most profound and fundamental questions human beings can ask: Who are we? Where did we come from? What is our future?

Astronomy, then, is the study of the origin, evolution, composition, size, and motion of the Universe as a whole and of the bodies in it: planets, moons, stars, galaxies, meteorites, asteroids, comets, and many others.

Astronomy and astrology

People often confuse the science of astronomy with the pseudoscience of astrology. In ancient Mesopotamia, where astrology is thought to have originated about five thousand years ago, astronomy and astrology were intimately bound together as people observed the stars and motions of the planets, Sun, and moon to try to predict the supposed influence of these objects on human lives.

Astronomy and astrology continued as partners for several millennia in other ancient cultures, including Egyptian, Chinese, Greek, and Roman through the European Middle Ages. Even such an illustrious astronomer as Johannes Kepler (1571-1630) was both astrologer and astronomer. Credit should thus be given to astrology for making possible many of the observations that have led to our modern science of astronomy.

The two words also have the same Greek root, *astr*, meaning star. The difference between astronomy and astrology, however, is found in the last part of the words. *Nomos*, as in astronomy, means law; astronomy is the study of the laws of the stars. *Logos*, as in astrology, means speech or discourse, thus giving the word astrology a meaning close to "told by the stars."

It is only in more recent times that astronomy and astrology have parted company. The reason for this split is that the accuracy of observation that the science of astronomy employs conclusively demonstrates that, with the exception of the tides caused by the gravitational pull of the moon and Sun, there is absolutely no influence on human affairs as a result of the movements and positions of celestial objects.

Astronomy seeks to uncover precise cause-effect relationships between observed phenomena and to arrange these observations into a total world picture of the structure, motions, and evolution of the Universe and all of its parts. Astronomers do not claim to be able to answer all the questions that we can ask about the Universe, or to be able to explain many of the things that happen in the Universe or on the Earth. They do not pretend to have absolute and complete knowledge about any part of the cosmos.

Astronomers do, however, work very carefully to gather data that can be repeated in subsequent observations, with the prospect that we will continue to learn more and more about the Universe. This kind of careful and painstaking accuracy is missing from astrology and other enterprises that have no basis in observational evidence and fact.

Science, including astronomy, is the procedure by which we try to understand nature and its operations. Astronomers pose hypotheses, for example, about how a star generates light energy. As we test these ideas against observed data, our hypotheses change until we have a workable model called a theory. Theories, too, change when new evidence is found.

Taking another illustration, planetary astronomers thought for many years that the seasonal color changes on Mars, observed through Earth-based telescopes, might be caused by growing vegetation watered by moisture from melting Martian polar caps. With the evidence gathered by the unmanned spacecraft, Mariner and Viking, that have orbited and landed on Mars during the past 25 years, it is now generally agreed that these changes are due to large and vigorous dust storms that occur periodically in the atmosphere of the planet. This new information altered our theories about Mars.

Dr. Bart Bok, one of the preeminent astronomers of this century, felt strongly that astrology, as important as it was to ancient peoples, should be regarded only as a historical curiosity that no longer has relevance for educated and well-informed people. He has written, "I have had more than half a century of day-to-day and night-to-night contacts with the starry heavens, and it is my duty to speak up and to state clearly that I see no evidence that the stars and planets influence or control our personal lives and that I have found much evidence to the contrary. Before the days of modern astronomy, it made sense to look into possible justifications for astrological beliefs, but it is silly to do so now that we have a fair picture of man's place in the Universe."[1]

1. "A Critical Look at Astrology," *The Humanist*, 35 (September/October 1975): 6-7.

Atoms

All matter and objects in the Universe are made of atoms, parts of atoms, or clumps of atoms. An atom has two basic components, a nucleus that is composed of protons and neutrons and a cloud of electrons orbiting the nucleus. Protons have positive electrical charges, electrons have negative electrical charges, and neutrons have no electrical charge. Protons and neutrons have about the same mass, approximately 1,836 times the mass of an electron. In a normal atom, the number of protons always equals the number of electrons. Thus, a normal atom itself has no electrical charge.

Different elements have different numbers of protons (and hence electrons). For example, hydrogen, the most abundant element in the Universe, has one proton in its nucleus and one electron orbiting the proton. Hydrogen is thus also the lightest atom. As seen in the diagram, helium, the second lightest and the second most abundant atom in the Universe, has two protons in its nucleus that are accompanied by two neutrons in the most common

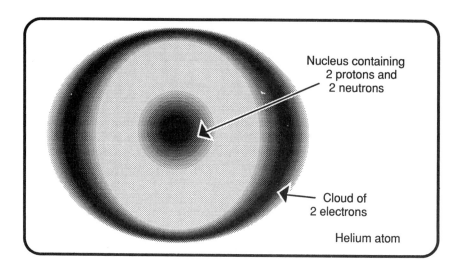

form of helium. There are two electrons orbiting the helium nucleus. Helium is the same element that we use to fill balloons. Since it is lighter than air, which is mostly oxygen and nitrogen, the balloons rise.

Hydrogen and helium together make up about 99 percent of all the matter in the Universe. All of the other 90 naturally occurring elements together comprise about one percent of the matter in the Universe. Other abundant elements in the Universe are:

Element	Protons	Neutrons
Carbon	6	6
Nitrogen	7	7
Oxygen	8	8
Neon	10	10
Magnesium	12	12
Silicon	14	14
Sulfur	16	16
Iron	26	30

Uranium is the most massive naturally occurring element, with 92 protons. Its most abundant isotope has 146 neutrons in the nucleus. Thus, we have $^{238}_{92}U$. Less abundant isotopes of uranium are $^{235}_{92}U$ and $^{234}_{92}U$.

Isotopes are variations of the same element due to varying numbers of neutrons in the element's nucleus. For example, if we add a neutron to the one proton in the nucleus of a hydrogen atom (of course, we do not add another electron since a neutron is electrically neutral), we have a heavier isotope of hydrogen called deuterium. If we add two neutrons to the hydrogen nucleus, we have an even heavier isotope of hydrogen called tritium.

In the foregoing illustration, the electrons are in their lowest energy level closest to the nucleus. If the helium atom absorbs energy, one or both of the electrons can move into higher orbits, or energy states. The atom is then said to be excited.

When an electron moves closer to the nucleus from a more distant orbit—when it moves from a higher energy level to a lower energy level—the atom emits a photon, or wavelength of energy. Some of this energy we see as visible light. Other energy is in the form of infrared or ultraviolet light, or even radio energy, x-rays, or gamma rays.

If an atom absorbs enough energy, an electron can be knocked completely away from its nucleus, and we have an ionized atom that obviously has a positive electrical charge (since an electron is no longer associated with the nucleus, there are more positively charged protons than negatively charged electrons associated with the atom). In the case of more massive atoms, oxygen, nitrogen, and iron, for example, the atom can absorb so much energy that it loses more than one electron.

If a hydrogen atom is ionized, the result is a free electron (not associated with a nucleus) and a free proton. If helium is singly ionized, the result is a free electron and a positively charged atom of helium, called an ion. If helium is doubly ionized, the result is two free electrons and a helium nucleus, also called an ion, or an alpha particle.

Atoms can also clump together to form molecules. A familiar molecule is water, H_2O. This symbol simply means that a molecule of water is made of two atoms of hydrogen and one atom of oxygen. The natural gas that we burn in our furnaces and stoves is another familiar molecule, CH_4, made of one atom of carbon and four atoms of hydrogen.

Aurora

When we are away from city lights or other distracting forms of light pollution, we may have the opportunity to see the aurora, particularly if we are in the northern part of the United States or in Canada. We are awed by the shimmering, shifting, sometimes rapidly moving colors of the *aurora borealis*, "the northern lights," in the nighttime sky. (In the Southern Hemisphere the aurora are called *aurora australis*, "the southern lights.")

These spectacular lights originate far from the Earth on the surface of the Sun. Solar flares and other high energy events on the Sun eject fast moving electrons, protons, and ions, together termed the solar wind, into the Solar System. Some of these charged particles are captured by the Earth's magnetic field.

These particles move along the lines of force of the Earth's magnetic field toward the North and South magnetic poles. Low-energy electrons collide with atoms and molecules in the Earth's atmosphere above our planet's North and South poles. Most of these collisions occur 80 to 160 kilometers (50 to 100 miles) above the surface of the Earth at high geographical latitudes.

The low-energy electrons excite or even ionize the oxygen and nitrogen atoms and molecules in our atmosphere. When these atoms and molecules return to their unexcited states, energy in the form of visible light—mostly reds, blues, and greens—is emitted and we see one of the most glorious sights the nighttime sky has to offer us. We can think of these atoms and molecules as fluorescing, much like the atoms of gas produce light in a fluorescent bulb.

These charged particles from the Sun are moving so fast that they travel the 93 million miles between the Sun and the Earth in an average time of 48 hours, meaning that they are moving at nearly 2 million miles per hour.

Since aurorae are caused by events at the surface of the Sun, and since the Sun reaches maximum solar activity every 11 years, we see more aurorae every 11 years during periods of maximum solar activity.

Big Bang and the expanding Universe

All of us have had the experience of watching and hearing a train approach a railroad crossing where we are standing. As the train roars toward us, its horn seems to have a higher pitch than it does when the engine has passed us; the horn then seems to drop to a lower note. This phenomenon, called the Doppler effect, is the result of the sound waves bunching together as the train approaches us and spreading out as the train recedes from us. The amount of change in the pitch of the horn, compared to the pitch of the train standing still, is proportional to the speed of the train. The faster the train approaches us, the higher the pitch; the faster the train recedes from us, the lower the pitch.

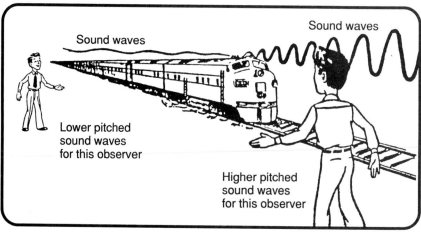

Sound waves

Sound waves

Lower pitched
sound waves
for this observer

Higher pitched
sound waves
for this observer

The same situation occurs with the light from stars and galaxies. If a galaxy is moving toward us, the light from that galaxy will be shifted toward shorter wavelengths, that is, toward the blue wavelengths of light compared to the wavelengths of the light we detect if the galaxy is stationary relative to us. Conversely, if a star or galaxy is moving away from us, the light from the object will be shifted toward the longer wavelengths, that is, toward the red wavelengths of light compared to a stationary star or galaxy.

The shifts in the wavelengths of light cannot be detected directly: to our eyes, the stars and galaxies moving toward or away from us do not change color. The shifts in wavelengths are observable, however, through a device called a spectrograph that is connected to a telescope. If galaxy A is moving away from the Milky Way, and thus the Sun and Earth, in photographs taken through the telescope and spectrograph, its light will appear to have shifted toward the red end of the spectrum.

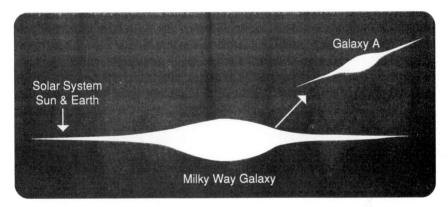

By the early 1930s, Edwin Hubble, an American astronomer working at the Mount Wilson Observatory in California, confirmed that most of the galaxies he studied have redshifts and are therefore moving away from us. He also discovered an even more important fact: the faster the galaxies are moving away from us, that is, the larger redshifts they have, the farther they are away from us. Thus, the ability to determine the speed of recession of a galaxy moving away from us, as measured by the redshift, gives us a way of measuring not only the vast distances of the Universe but also the size of the observable Universe to the limits of our largest telescopes.

If most of the galaxies are moving away from us, we have the impression that we must be at the center of the Universe. However, this is not the case. Imagine a loaf of raisin bread dough about to rise. The raisins represent galaxies, and the loaf is the Universe. As the loaf (Universe) expands, all the galaxies (raisins) move away from each other. No matter which galaxy (raisin) we are on, all the other galaxies (raisins) will move away from us. Thus, no matter where we are in the Universe, it appears as if we are at the center. In reality, of course, the Universe has no center. This situation of all galaxies moving away from each other we call the expanding Universe.

Big Bang and the expanding Universe: How it all began

Cosmology is a branch of astronomy that investigates how the Universe began, how it evolved to what we observe today, and where it may be going in the future. The Universe appears to be expanding because galaxies are moving away from each other. Thinking back into time, it would seem logical that the mass in the Universe that now appears to be expanding and thinning out in the form of galaxies at one time existed in a primeval fireball in which all the matter was densely packed together. Current theory suggests that this original fireball, composed of radiation, would have had a very high density and a very high temperature, on the order of many billions of degrees Kelvin.

When the fireball initially exploded, the Big Bang, the temperature dropped to one billion degrees in about three minutes and the particles began to form the two lightest atomic elements, hydrogen and helium. About 80 percent of the matter of the original fireball after the explosion was hydrogen, and 20 percent of the matter was helium, according to the most commonly accepted theory.

As the fireball continued to expand for the next million years or so, the temperature dropped to a few thousand degrees and the density to several thousand atoms per cubic centimeter, a very low density compared to gases with which we are familiar. After perhaps a billion years, theory predicts that the hydrogen and helium would be ready to condense into stars and galaxies, a process that apparently is still taking place today.

We must remember that the Big Bang did not occur in one place because the Universe has no center. The explosion thus occurred everywhere without boundaries. What we see today is thought to be the continuing expansion of the original event.

If the Universe began with a bang, then it must have a finite age. No one knows, of course, what the precise age of the Universe is, but the best estimates at this time are between 13 billion and 20 billion years.

When galaxies and other objects are photographed through large telescopes, the most distant objects look to us as they might have appeared near the beginning of the Universe. This situation occurs because the speed of light is finite—300 thousand kilometers per second (186 thousand miles per second). For example, because the Sun is about 150 million kilometers away from the Earth, we do not see it as it appears at the instant we look at it. We see it as it was $8\frac{1}{3}$ minutes ago because it takes light traveling at 300 thousand kilometers per second that amount of time to reach the Earth.

In a similar way, when we photograph a distant galaxy or other object that may be several billion light-years away, we record light that left the galaxy several billion years ago. Therefore, we photograph distant galaxies and other objects not as they appear now (we have no way of knowing what they look like now, or even if they still exist), but as they appeared many billions of years ago, perhaps close to the time they and the Universe itself came into being. Thus, whenever we look out into space, we unavoidably look back into time.

Big Bang and the expanding Universe: The future

There are essentially two possible futures for the Universe. It might be open, which means that it will expand forever: galaxies will continue to move away from one another; density of matter will become thinner; and the Universe will never again, even in an infinite amount of time, return to any previous condition.

The other future for the Universe is what the astronomer describes as closed. The Universe will continue to expand to some maximum size, at which point gravity will overcome the energy of expansion left over from the Big Bang, and all the galaxies will begin moving toward one another. Presumably the galaxies would recongregate, and the matter in the Universe would again form the primeval fireball. If the Universe is closed, the times at which it would reach maximum expansion and final collapse are unknown.

A variation of the closed situation is the oscillating Universe. Some astronomers have suggested that, if the Universe does collapse into a fireball again, it may go through another Big Bang. Matter would repeat its star and galaxy formation and expansion. If this is the case, no time span for the cycle, or oscillation, is known.

There are several critical tests that astronomers are carrying out now to determine whether the Universe is open, closed, or oscillating. One of these tests involves a determination of the amount of mass in the Universe. If there is enough mass in the Universe to create a field of gravity that has the strength to slow expansion, eventually the Universe will stop expanding and begin to collapse. Preliminary results from the data indicate that there is not enough mass to stop expansion, and that the Universe may be open. It is emphasized, however, that there is insufficient evidence at present to demonstrate conclusively either an open or a closed Universe.

Another test has to do with how fast the expansion of the Universe is slowing. We know that the Universe was expanding faster earlier in its history than it is now. This conclusion is based on the observation that more distant galaxies, i.e., those galaxies that we observe in earlier periods in the history of the Universe, are moving away from us faster than galaxies that are nearby in space and thus closer to us in time. The only possible conclusion is that the general expansion of the Universe is slowing.

The question is: How fast is the Universe slowing? If the Universe is slowing fast enough, it will eventually stop expanding; that is, the Universe is closed. If the Universe is not slowing fast enough, it will continue to expand forever (even though the rate of expansion always slows).

Whatever the ultimate disposition of the Universe, we here on Earth need have no concern about our future. The time scales are so large that there will be no effect on life on our planet. Human curiosity and intellect, however, will keep searching for answers to these truly profound questions about the future of our Universe.

Black holes

Stars shine because atoms of hydrogen, the lightest and most abundant element in the Universe, are fused together in the stars' hot, dense interiors. Elements such as helium, carbon, and oxygen are created at the same time that energy is released. We see this energy as visible light or record it on photographic film with a camera or other detectors attached to a telescope.

Suppose we start with a star that contains several times as much matter as the Sun. (There are stars that contain 10 or even 80 times more matter than the Sun.) When some of the hydrogen atoms have been fused into other elements, the star will collapse, due to gravity, into a smaller, hotter, and more dense star. It then explodes, scattering matter in all directions into space. The core that is left behind, if it continues to collapse, may form a black hole. The matter in this super dense core is so closely packed together that one tea-spoonful may weigh millions of tons.

We can thus think of a black hole as starting with a star that is a great deal more massive than the Sun. The matter in this massive star is compressed by gravity into a sphere only a few kilometers in diameter.

Black holes are so named because they cannot be seen. No light—or anything else—can escape from them. We see the Sun because visible light, although it has to do work to leave the Sun, can escape because our star does not have a strong enough gravitational field to stop it. A black hole, however, has an incredibly strong gravitational field because so much matter is squeezed into a very small volume. Light cannot do enough work—it is not energetic enough—to escape from a black hole. We may say that the black hole has disappeared from the Universe.

Even though we are not able to observe black holes directly, astronomers think they can be detected. They have such large gravitational fields that nearby matter would be sucked into a black hole. Radiation in the form of x-rays may be produced as this matter falls into the hole; several strong x-ray sources that may turn out to be black holes have been found by scientific satellites.

It has been speculated that matter inside black holes is squeezed together to the extent that it may become dimensionless, a singularity having no height, width, or depth. Albert Einstein and others speculated, in the 1930s, that the matter from black holes pops through them into what have been called white holes someplace else in our Universe or perhaps makes its appearance in someone else's Universe.

No one knows, of course, what happens to matter inside black holes. Astronomers are just now beginning to accumulate evidence that black holes exist. If there really are black holes, they may turn out to be some of the most puzzling and exciting objects yet discovered in the astronomical Universe.

Tycho Brahe

Tycho, as he is more familiarly known, was born into a noble family in 1546 in what was then a part of Denmark. He died in 1601, eight years before Galileo invented the first astronomical telescope.

Generally regarded as one of the greatest pre-telescopic observers, Tycho spent his eventful and energetic life observing the positions of stars and the motions of the planets. In 1575, King Frederick II of Denmark gave him the Island of Hveen, near Copenhagen, together with the support staff and money necessary to build an elaborate observatory. It has been suggested that these gifts comprise the largest grant ever given to one person for scientific research.

In the days before telescopes, observing equipment was basically sighting instruments, such as the quadrant, that were used to measure the positions of stars and planets with respect to the horizon. Tycho designed and built his own observatory instruments, more accurate than any that had been built previously.

Further, Tycho's data were more accurate and more complete than any data prior to his work. Tycho did little to interpret his observations. His contribution to our understanding of the Universe was to provide the facts that could be used by other astronomers, principally his brilliant assistant, Johannes Kepler.

Stellar parallax is the illusion we see that nearby stars shift slightly against the background of more distant stars as the Earth revolves around the Sun. Because Tycho could not observe stellar parallax, he concluded that the Earth must be motionless in the center of the Universe, as Ptolemy and earlier Greeks had taught. Thus, Tycho never adopted the heliocentric (Sun-centered) interpretation of the Universe proposed by Copernicus in 1543. Today we know that even the nearest stars are so far away that their parallax can be observed only with sophisticated telescopes, and not at all by the unaided eye as Tycho was trying to do.

Up to Tycho's time, Aristotle's explanation that comets were part of the Earth's atmosphere was accepted by almost everyone. Tycho showed, however, that the comet of 1577 displayed no parallax when viewed against the background of the more distant stars from several places on the Earth. Had there been an observable parallax, the comet would have been quite close to the Earth and thus in the Earth's atmosphere. Because the comet showed no parallax, it must be in the realm of the planets.

Aristotle further suggested that this region of the Universe was immutable. Yet, comets clearly changed shape and brightness as they moved through the supposedly immutable realm of the planets. Furthermore, they had to move through the crystalline spheres that were supposed to carry the planets, a difficult feat to imagine. Thus, Tycho supplied more evidence that would eventually help to destroy the old Greek ideas of the Universe.

In 1572, there occurred a brilliant supernova, an exploding star that was meticulously described by Tycho. Again, the immutable, perfect Universe assumed by Aristotle and other Greek thinkers proved to have exceptions.

Tycho can truthfully be said to be one of the founders of modern science, together with Copernicus, Kepler, Galileo, and Newton, all of whom began to shape our contemporary ideas of the Universe and to prescribe our modern scientific methods of data acquisition and analysis.

Comets

Comets have been observed ever since human beings speculated about objects in the nighttime sky and appear in records from the beginning of recorded civilization.

Almost certainly comets are members of the Solar System. There may be a large cloud of cometary material, perhaps debris left over from the origin of the Sun and planets, far beyond the orbit of Pluto, the planet that travels farthest from the Sun. Occasionally, some of this material will be pulled toward the Sun, and if there is enough material as it approaches the Sun, it will be visible to our unaided eye as a comet. It is thought that there may be as many as one hundred billion comets in the cloud with a combined mass equal to that of the Earth.

There are accurate orbits calculated for about one thousand comets. Some of these comets approach the Sun only once; others, called periodic comets, make repeated trips around the Sun. The shortest period for a comet is $3\frac{1}{3}$ years, and the longest known period is about 151 years.

The most famous and spectacular periodic comet, at least in the past few thousand years, is Halley's comet (Halley rhymes with valley), last visible in 1985-1986. Since it takes the comet about 76 Earth years to make one trip around the Sun, we should see it again in the year 2061.

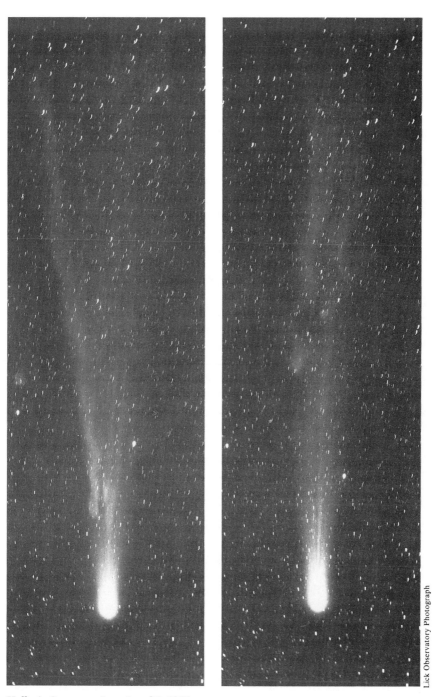

Halley's Comet on June 6 and 7, 1910.

Lick Observatory Photograph

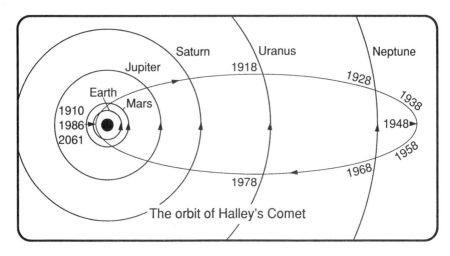

The orbit of Halley's Comet

Comets have a nucleus that, in the largest ones, is 10 kilometers in diameter. The nucleus is composed of frozen gases. (The temperature where comets spend most of their lives far from the Sun is perhaps less than four hundred degrees below zero Fahrenheit.) Imbedded in the frozen gases are interplanetary dust and tiny fragments of stony and metallic meteoric material. The frozen gases are mostly water vapor together with lesser amounts of carbon dioxide and carbon monoxide. Ammonia, methane, and other carbon compounds are also present, as well as molecular hydrogen.

Around the nucleus is the coma that may be up to one hundred thousand kilometers in diameter. The coma is composed of atoms and molecules of gas that travel with the nucleus. The coma and nucleus together are called the head of the comet.

As a comet approaches the Sun, the gas becomes unfrozen, thus liberating the solid particles of dust and meteoric material. Among other forces, light pressure from the Sun pushes these materials away from the head and in a direction opposite the Sun. Comets thus develop two tails as they approach the Sun. One tail is made up of dust, and the other is gas, often called the ion tail. One or both of the tails can be up to 150 million kilometers long.

We see comet heads and tails because they not only reflect sunlight, as planets do, but the atoms of the gases also fluoresce—give off their own light—when comets are near the Sun. Because periodic comets lose material every time they orbit close to the Sun, it has been calculated that even a large comet has enough material to make only a few thousand trips around the Sun.

Both the coma and tail of a comet contain so little material that they rival the best laboratory vacuums. The Earth apparently passed through the tail of Halley's comet, back in 1910, with no noticeable effects due to the low density of material in the comet's tail. Some people predicted the end of the world at that time, and others made money selling anti-comet pills—whatever they are. We can safely say that nothing happened to the planet Earth.

Constellations

Why did no one teach me the
constellations when I was a child?
—Thomas Carlyle

Ancient people organized the maze of stars that can be seen in the nighttime sky into groups of familiar objects. These objects reflected their preoccupation with hunting activities—for example, there is Orion, the Mighty Hunter, followed by his faithful dog, Canis Major; Leo, the Lion; Ursa Major and Ursa Minor, the Big and Little bears; and Vulpecula, the Fox.

Farming people and animals are also popular: we have Taurus, the Bull; Aquarius, the Water Carrier; Aries, the Ram; and Boötes, the Herdsman. Characters from mythology, such as Cepheus and Cassiopeia, the King and Queen of Ethiopia, and Andromeda, the Princess of Ethiopia who was rescued by Perseus, are also important. Commonplace animals are also seen in several groups of stars: Cancer, the Crab; Cygnus, the Swan; Lepus, the Hare; Pisces, the Fish; and Scorpio, the Scorpion. All of these constellations can be seen from the Northern Hemisphere.

It is interesting to note that many of the southern constellations, however, are named after more modern, technological objects: Microscopium, Telescopium, Sextans and Octans (Sextant and Octant), Pictor (Painter), Antlia (Air Pump), Fornax (Furnace), and Carina and Puppis (the Keel and Stern of the Argonauts' Ship). The reason for modern designations is that much of the Southern Hemisphere was uninhabited until recent historical times, or it was inhabited by peoples whose star pictures did not make their way into our star lore.

The word constellation means studded with stars. Now, there are 88 recognized constellations, with every star in the sky contained in a constellation. The oldest names cited above are Latin translations of Greek names. The Greeks, however, appropriated many star group names from older civilizations such as the Sumerians, Babylonians, and people living in the Euphrates River valley.

We are all familiar with the Big Dipper of the northern sky. This constellation was also called Ursa Major, the Big Bear, by many different ancient peoples, thus indicating a great deal of cultural cross-fertilization in past times. It is curious that the Bear was usually depicted as having a long tail—the handle of our dipper—when all existing species of bears have short, inconspicuous tails. No one has yet explained this discrepancy satisfactorily.

Astronomers still use constellations to locate objects in their approximate positions in the sky. For example, we may say that Mars is in Pisces; this simply means that, from the Earth, Mars appears against the background of stars called Pisces. There is no physical connection between Mars and the stars; indeed, there is often no physical connection between the stars themselves in a constellation. They only appear from the Earth as if they are arranged in a group that has a physical relationship.

Why should we learn the names of the constellations today? One answer has to do with familiarity. We all feel more comfortable and at home in a forest if we can identify some of the trees and animals around us. Fear and puzzlement are reduced in an environment when we recognize familiar elements in that environment. Stars, too, are a part of our environment. Knowing the constellations, their names, the stories that go with them, and the names of the brightest stars in the constellations gives us the sense of familiarity with old friends and, hence, puts us into a more comfortable relationship with our entire environment.

Copernicus

For most of history, people believed that the Earth is the center of the Universe. Nicholas Copernicus, born in Poland in 1473, spent his life considering an alternative explanation, one that would also account for the apparent daily motions of the Sun and stars across the sky and for the apparent motions of the Sun and planets against the background of the fixed stars.

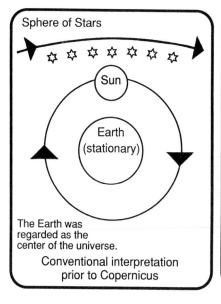

Sphere of Stars

Sun

Earth
(stationary)

The Earth was regarded as the center of the universe.

Conventional interpretation
prior to Copernicus

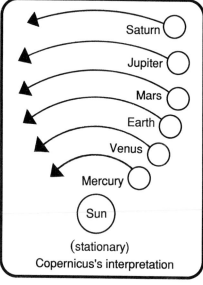

Saturn

Jupiter

Mars

Earth

Venus

Mercury

Sun

(stationary)

Copernicus's interpretation

Copernicus argued in his major book, *On the Revolutions of the Celestial Spheres*, published the year he died (1543), that the Sun should be seen as stationary, and that the Earth, the other planets, and the stars can be interpreted as revolving around the Sun (see drawing p. 24). Copernicus further suggested that the Earth was spherical and that it rotated on its axis; this explanation would account for the daily rising and setting of the Sun and fixed stars:

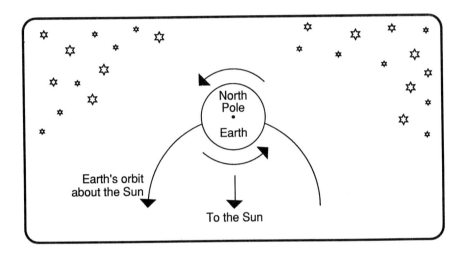

In his monumental book, he also correctly laid out the general scheme of the Solar System, noting that Mercury was closest to the Sun, followed by Venus, Earth, Mars, Jupiter, and Saturn. (Only the naked-eye planets, together with the Earth, were known in Copernicus's time.) He also deduced that the closer a planet is to the Sun, the faster it has to move around the Sun. Thus, Mercury revolves around the Sun in about 88 days, Venus in 225 days, and the Earth in 365 days. Copernicus even worked out an approximately correct scale for the distances of the then-known planets from the Sun.

The older Earth-centered system was very complicated, particularly when it attempted to account for the motions of the planets. Copernicus's system was much easier to understand; he thus seems to have been motivated to describe a Sun-centered Solar System because it was simpler and more aesthetically appealing.

While Copernicus never attempted to prove that the Earth revolved around the Sun, he did suggest that the Earth was similar to the other planets in that it is an astronomical object. His ideas were an enormous step forward in our understanding of the Universe because they moved human thinking from an Earth-centered to a Sun-centered Universe and, thus, laid the foundation for modern astronomical science.

The Coriolis effect

Suppose we shoot a cannon ball from the North Pole directly south toward the equator. If the Earth were not rotating, the cannon ball would land directly south at position 1, as illustrated in Diagram A. The Earth does rotate,

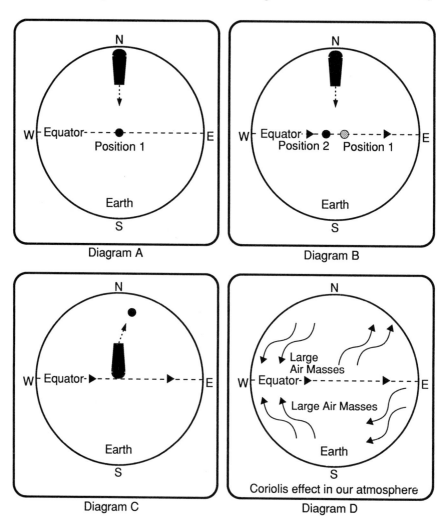

Diagram A

Diagram B

Diagram C

Coriolis effect in our atmosphere

Diagram D

however, from west to east. While the cannon ball is in flight toward the equator, the point on the equator at which the cannon was aimed has moved toward the east. The cannon ball will land at position 2 (see Diagram B) and, thus, seem to be deflected toward the right as seen from the North Pole.

If we move our cannon to the equator and shoot it directly north, the cannon ball will again seem to be deflected to the right (see Diagram C). The apparent deflection of cannon balls, rockets, and similar objects is called the Coriolis effect and was first analyzed in 1835 by Gaspard Coriolis. The Coriolis effect is a proof of the rotation of the Earth on its axis.

Not only do cannon balls appear to be deflected, but large air masses also experience the apparent deflection. In the Northern Hemisphere, large wind patterns cycle counterclockwise, and in the Southern Hemisphere, they cycle clockwise (see Diagram D).

The Coriolis effect is thus responsible for large-scale movements of air in the atmosphere of our planet. We can also use our knowledge of the Coriolis effect to analyze atmospheric phenomena such as zones, belts, the Great Red Spot, and other spots in the atmosphere of Jupiter, as well as the atmospheric phenomena of the other giant gas planets: Saturn, Uranus, and Neptune.

Returning to the Earth for a moment, if we had a perfectly symmetrical bathtub, basin, or toilet bowl and let the water run out, in the Northern Hemisphere, it would run out in a counterclockwise swirl. In the Southern Hemisphere, the water would run out clockwise. Exactly on the equator, the water would run straight down.

Cosmic microwave radiation

In 1965, two scientists at the Bell Laboratories in New Jersey, while working on another project, discovered radiation that came from every direction in the sky. What they discovered was the faint energy left over from just after the Big Bang. This cosmic microwave radiation is the most distant and hence oldest signal that we have received from the early Universe. In other words, it is farther back in the history of the Universe than any other signal we have received.

Early in the history of the Universe, just a few hundred thousand years after the Big Bang when the Universe still had a temperature of several thousand degrees, it was filled with high-energy, short-wavelength radiation, the result of the high temperature. The Universe is now thought to be on the order of 13 billion to 20 billion years old. During this vast amount of time, it has continuously expanded. The original temperature of several thousand degrees has now dropped to just about three degrees Kelvin above absolute zero. The high-energy, short-wavelength radiation has now become low-energy, long-wavelength radiation, precisely the radiation discovered in 1965. We properly call this energy that saturates the Universe the 3-K cosmic microwave background radiation. The current radiation has a wavelength of about one-tenth of a centimeter in the microwave part of the radiative energy spectrum. These are the same wavelengths that we use to cook with in microwave ovens. We will not be able to cook with the cosmic microwave radiation, however, because the energy is so diffuse; in microwave ovens we concentrate the microwaves many times more than the cosmic microwave radiation.

The presence of this radiation gives us several clues about the early Universe. It shows us that there was indeed a Big Bang because there is no other way to explain the background microwave radiation.

Another clue to the early Universe carried by the cosmic microwave radiation is that its intensity is isotropic, the same in all directions. Careful measurements since the discovery of the cosmic microwave radiation in 1965, including those observations made by the Cosmic Background Explorer (COBE) satellite launched in 1989, indicate that there is no variation in intensity anywhere in the sky. In turn, these observations mean that the early Universe was smooth: it had no clumps or lumpiness of matter. If the early Universe were as smooth as the cosmic microwave radiation suggests, then we still do not have an explanation of how the lumps of matter that we now call galaxies could have originated.

The only exception to the lack of variation in the intensity of the cosmic background radiation is that it is slightly more intense in one direction and less intense in exactly the opposite direction. This fact reflects the motion of the Earth through the "sea" of the cosmic background radiation. If we subtract the motion of the Sun around the center of the Milky Way Galaxy, which of course is taking the Earth and the rest of the planets and Solar System with it, we find that our entire galaxy, together with the Local Group of galaxies, is moving away from the Virgo cluster of galaxies, some 60 million light-years away, slower than it should in the general expansion of the Universe. Our movement through the cosmic microwave radiation thus suggests that there must be a supermassive concentration of galaxies in the direction of Virgo whose gravitational field is slowing the expansion of the Universe, at least locally (if we can think of 60 million light-years as being local).

Cosmic rays

Every second, 1,000,000,000,000,000,000 particles, called cosmic rays, enter the Earth's atmosphere at nearly the speed of light from all directions in space. (We have no name for a number this large; in order to save space, we write it simply as 10^{18}, which means a one followed by 18 zeroes.) The Earth receives more energy from cosmic rays than we receive from the light of all the stars combined.

Cosmic rays come in two varieties. The first are called primary cosmic rays because they have travelled over interstellar and perhaps even intergalactic space from their sources. When they enter the Earth's atmosphere many of them collide with molecules of gases in our atmosphere. These collisions produce the other variety, called secondary cosmic rays. These secondary rays in turn collide with other molecules, and we have a shower of cosmic rays.

Most primary cosmic rays are high-speed nuclei of hydrogen atoms, called protons; many of them are also the nuclei of helium atoms, called alpha particles. Less than one percent of cosmic rays are the nuclei of heavier atoms. These nuclei include iron and even a few nuclei of more massive atoms.

It is probably fortunate that we cannot see or feel cosmic rays. If we could, we would be dazzled by the enormous and constant bombardment by cosmic ray energy of our atmosphere and everything around us, including our own bodies. A few cosmic rays, of course, collide with atoms in our bodies, as well as with atoms in the tissue of other living plants and animals. It is these collisions of cosmic rays with material inside the cells of living organisms that scientists think are one of the causes of genetic mutations and hence a source of variation in biological evolution. Thus, cosmic rays influence the way life develops here on Earth. Cosmic rays may also be one of the natural causes of cancer in human beings.

The origin of cosmic rays is not yet well understood. It is known, however, that within the Solar System only the Sun can produce cosmic rays, but it accounts for only a small number of them. Most cosmic rays we detect probably originate within our Milky Way Galaxy, perhaps as the result of explosions of giant stars called supernovae. A few of the very high-energy cosmic rays may have originated in galaxies beyond the Milky Way and, thus, have travelled through space for millions or even billions of years before they reach us. We call these particles cosmic rays because they come to us from all regions of the cosmos.

The Cosmological Principle

Imagine that you are in a jet airplane flying at 35 thousand feet. You get up from your seat to walk to the front of the plane. What is your speed?

The answer to this question depends on the frame of reference in which you choose to measure your motion. For example, with reference to the other passengers in the airplane, you are moving about 1 kilometer per hour. Yet, with reference to the center of the Earth, you are moving at the speed of the plane, let us say 600 kilometers per hour, plus your walking speed, or 601 kilometers per hour.

What is your speed with reference to the ground? If the plane is flying east to west, you have to take into consideration your walking speed, plus the speed of the plane relative to the center of the Earth, plus the speed of the Earth that is rotating under the plane from west to east.

But there is more. The Earth revolves around the Sun at a speed of about 30 kilometers per second with reference to the Sun. The Sun itself is moving around the center of our galaxy at a speed of about 220 kilometers per second. Our galaxy is moving at a speed of hundreds of kilometers per second with reference to nearby clusters of galaxies, and at speeds of tens of thousands of kilometers per second with reference to distant galaxies. Thus, all objects in the Universe are in motion relative to all other objects, whether these objects be airplanes, passengers, stars, planets, galaxies, or clusters of galaxies.

Therefore, in order to answer the question about your speed as you walk down the aisle of an airplane, you have to specify your frame of reference. Put another way, the Universe has no preferred or absolute frame of reference. All motion is relative to whatever frame of reference we specify.

A little thought will show us that if there is no absolute frame of reference, all parts of the Universe, over large enough distances, will look the same. That is, if there were a preferred or absolute frame of reference, it would look different from the other parts of the Universe; it would be unique. Since, however, we have just demonstrated that there is no unique, absolute, preferred frame of reference in the Universe, every direction in the Universe should look exactly like every other direction. Thus, when we look out at galaxies in any direction we see them fleeing away from us, as measured by their recessional velocities, as the Universe expands. In other words, the Universe is the same in all directions; we say the Universe is isotropic.

Furthermore, over large enough distances and volumes, the Universe has an even distribution of matter and is said to be homogeneous. Put another way, all lumpiness disappears on large enough scales, and the Universe becomes smooth.

A good analogy is the page on which you are reading these words. On the large scale that you observe the page, it appears to be perfectly smooth. If you could see the page on the scale of molecules and atoms, however, it would indeed appear to be lumpy.

These two concepts, isotropy (the same in all directions) and homogeneity (even distribution of matter) together are called the Cosmological Principle.

No one is certain at what scale the Universe smooths out, but there are large structures composed of clusters and superclusters of galaxies that appear to be hundreds of millions of light-years across. (One light-year equals 9½ million, million kilometers.) Thus, the scale at which the Universe becomes smooth must be at least hundreds of millions of light-years in extent.

Another way of stating the idea that the Universe has no unique, absolute, or preferred frame of reference is to suggest that the Universe has no center. If the Universe had a center, that would be a unique place, but our discussion of the passenger's speed walking down the aisle of an airplane shows that there is no absolute speed because there is no absolute, unique, or preferred frame of reference against which to measure the passenger's speed. If the Universe has no center, then neither can it have an edge. An edge would be another unique, absolute frame of reference.

We can state these ideas one more way. The Universe has as many centers as there are places such as planets and galaxies from which to observe them. In the same way, the Universe has as many edges as there are places from which to observe them.

Finally, as the Universe expands—and taking into consideration the Cosmological Principle—it cannot be expanding into anything, such as a larger space, because that larger space would have to be regarded as a unique, absolute, and preferred frame of reference. Thus, as the matter in the Universe expands, as the clusters of galaxies speed away from each other, they create the space into which they expand.

It is indeed a strange and wonderful Universe in which we live.

Dark matter

Most galaxies exist in larger clumps of matter called clusters. Galaxies in a cluster are not only gravitationally bound together, but they are also moving with respect to each other within their cluster.

In the 1930s, it was discovered that many galaxies are moving too fast in their cluster to remain members of their cluster. There is just not enough visible mass in the form of stars in a cluster of galaxies to provide the gravitational strength needed to hold the cluster together. And yet, we observe clusters of galaxies holding onto their member galaxies.

In the 1970s and 1980s, the rotational speeds of many spiral galaxies were measured. Since galaxies are not solid, rigid objects like wagon wheels, they can rotate differentially; that is, parts of galaxies next to each other do not always stay in the same positions. Careful observation and analysis show that many spiral galaxies do not rotate from their centers through their arms at the speeds they should based on their visible mass.

Hence, we have two firm pieces of evidence that there is more matter in the Universe than meets the eye. Our evidence for this matter is the indirect, gravitational influence on visible matter. No one has "seen" this matter in visible light, radio, ultra-violet, infrared, or any other wavelengths. It is therefore referred to as dark matter. (It is sometimes erroneously referred to as the missing matter in the Universe, but we know it is not missing because we can observe its gravitational influence on visible matter, as noted.)

What can this dark matter be? There are many ideas, but no confirmed theories. Some astronomers suggest that it might be neutrinos: tiny, elusive particles that have no electrical charge. No one is even certain that they have mass. If they do, each neutrino would carry with it only a tiny amount of mass. Since neutrinos are the most common particle in the Universe, however, their great numbers could add up to a significant mass—enough, perhaps, to cause the gravitational effects that have been observed on rotating galaxies and clusters of galaxies.

A second explanation is that the dark matter may be in the form of innumerable dim stars associated with galaxies. These stars are so faint that they escape detection by modern telescopes, even in the comparatively nearby regions of our own Milky Way Galaxy. We will have to await better instruments before we can test this idea.

It may be that the dark matter is contained in large numbers of black holes which could not be detected directly. Cold gas, giving off so little radiation that it has escaped our instruments, is another candidate for the dark matter. Or it may be that the dark matter is in the form of some exotic, hypothetical particles such as cosmic strings.

How much dark matter is there? One intriguing idea is that there may be enough dark matter to close the Universe. The amount of visible, luminous matter in the Universe that we have observed to date is far too little to stop its expansion. We know that the expansion rate of the Universe is slowing down, but current data indicate that the rate is not slowing enough to keep the Universe from expanding forever. We thus say the Universe is open. Some astronomers suggest, however, that there may be sufficient dark matter to create a strong enough gravitational field to slow the expansion of the Universe to cause it to stop expanding. That is, there may be enough dark matter to close the Universe.

Improved instrumentation will solve the mystery of the dark matter and tell us what the future of the Universe will be. Will the Universe expand forever, or is there enough dark matter to stop the expansion and start the Universe contracting toward a "big crunch"?

Days of the week: How they got their names

The length of a day is measured by the amount of time the Earth takes to rotate once on its axis. Another way of stating the length of a day is to say that two successive passages of the Sun across an imaginary line in the sky, called a meridian, make a solar day. By definition, there are 24 hours in a solar day.

Astronomers also recognize a sidereal day, defined as the amount of time it takes for two successive passages of the same star across the meridian. (The word sidereal means star.) Because the Earth is also revolving around the Sun, a star crosses the meridian about three minutes and 56 seconds sooner each day. We say that the solar day is thus almost four minutes longer than the sidereal day (see drawing p. 33).

Due to tidal friction and friction between the Earth's molten interior and the more solid mantle, the Earth's period of rotation is slowing down. The length of a day is thus increasing at an average rate of 2/1,000 of a second per century.

The names of the days in the week and the names of the seven objects in the sky that are not the fixed stars were the same to the ancient Romans.

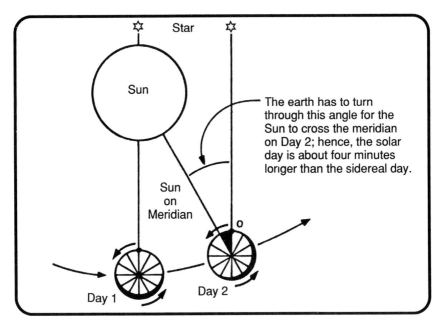

These objects are the Sun, the moon, and the five planets visible to the naked eye: Mercury, Venus, Mars, Jupiter, and Saturn.

Our Sunday, of course, is the day of the Sun; in Latin, it was named *Dies Solis*. Monday is the day of the moon.

Tuesday is more complicated. In Roman times, this day was *Dies Martis*, after Mars, the god of war. When we look at the planet Mars in the nighttime sky, it appears red to us, suggestive of blood and war. Our name comes from *Tyr* (in Saxon, *Tiw*), who was the god of war and justice to ancient Germanic peoples. *Dies Mercurri* to the Romans is our Wednesday, named after Woden (or Odin), the principal Scandinavian god. The Romans associated this day with the god and planet Mercury.

Thursday is named for the great German warrior-god, Thor. (Our word thunder is also derived from Thor.) To the Romans, this day was *Dies Jovis*, after their god, Jupiter. It is interesting to note that the planet Jupiter, named for the greatest Roman god, is also the largest planet in the Solar System; the Romans, of course, were unaware of its relative size.

Venus was the Roman goddess of charm and beauty. As the third brightest object in the sky, behind the Sun and moon, perhaps the planet of the same name also impressed the Romans as a beautiful and charming heavenly body. Our name Friday is taken from Venus's counterpart in Scandinavian mythology, Frigga, wife of Woden. Thus, in both Roman and Scandinavian mythology, this day is associated with an important female.

The association of Saturday with the planet Saturn is obvious. Saturn was the Roman god of agriculture.

Distances in the Universe

Calculating distances in the Universe is very important work for the astronomer because precise knowledge of distances gives us a clue about the size of the Universe, and hence about the length of time the Universe has been in existence.

Within the Solar System, astronomers reckon distances not only in miles or kilometers, but also in astronomical units (AU). One astronomical unit is simply the distance between the Earth and Sun, about 150 million kilometers (93 million miles). Astronomers often talk about the distances between the Sun and planets in astronomical units. For example, we speak of the distance of Mercury from the Sun as 0.39 AU (about 60 million kilometers or 36 million miles). Mars is 1.52 AU (about 227 million kilometers or 141 million miles) from the Sun. Pluto, the most distant planet, is 39.5 AU, or about 5.9 billion kilometers (3.7 billion miles).

Beyond the Solar System, distances are so large that astronomers use another unit of measurement called the light-year. Light travels at approximately 300,000 kilometers per second (186,000 miles per second). This is such a fast speed that a beam of light could travel around the Earth's equator, which is 40,250 kilometers or 25,000 miles in length, $7\frac{1}{2}$ times in one second!

A light-year is defined as the distance a beam of light travels in one year at the speed of 300,000 kilometers per second. The number of kilometers in a light-year can be calculated by multiplying 300,000 (the number of kilometers that light travels in one second) by 60 (the number of seconds in a minute) by 60 (the number of minutes in an hour) by 24 (the number of hours in a day) by 365 (the approximate number of days in a year). The answer is 9,460,000,000,000 kilometers (approximately six million, million miles). Now you know why astronomers do not use kilometers or miles to express distances in the Universe: the numbers simply are too large.

The distance from the Earth to the second closest star (the Sun is the closest star), called Proxima Centauri, is about $4\frac{1}{3}$ light-years, or 40,678,000,000,000 kilometers (26 million, million miles). In other words, to travel to Proxima Centauri, we would have to travel at 300,000 kilometers per second for more than four years. To travel to Sirius, the brightest star in the sky except for the Sun, we would have to travel at the speed of light for nine years. A trip to the North Star would take 820 years. For comparison: the Sun is only $8\frac{1}{3}$ light-minutes away.

Distances: How big is the Universe?

The light-year, the standard used by astronomers to measure distances in the Universe, is the distance light travels in one year: more than nine million, million kilometers (six million, million miles). To give an idea of the enormous distances between objects in the Universe, we need only consider the size of the Milky Way, the Galaxy of stars of which the Sun, Earth, and the rest of the Solar System are a part. The Milky Way is shaped like a giant pinwheel. If we were to start a beam of light at one edge of the Galaxy, that light, traveling at the speed of 300 thousand kilometers per second (186 thousand miles per second), would take more than one hundred thousand years to travel across our galaxy.

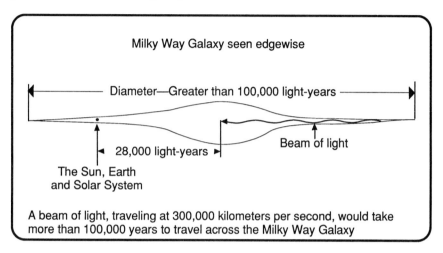

We thus say that our galaxy has a diameter of more than one hundred thousand light-years. If we want to calculate how many kilometers that is, we multiply 9.5 million, million (the number of kilometers in one light-year) by one hundred thousand; the answer is 950,000,000,000,000,000 kilometers. The Sun, Earth, and Solar System are located about 28 thousand light-years from the center of the Milky Way.

The next closest galaxy, the Andromeda Galaxy, is more than two million light-years from the Milky Way. Light that we see from this galaxy has been traveling toward us at three hundred thousand kilometers per second for more than two million years, or longer than there have been human beings on the Earth.

No one knows—yet—how big the Universe is. The most distant galaxies, those that can just barely be recorded with the largest telescopes, are probably several billion light-years away. Quasars, perhaps the most distant objects in the Universe, may be even farther away, close to the edge of our observable Universe.

Even though we cannot be precise about the size of the Universe, we can, with confidence, say that it is at least several billion light-years across. Perhaps larger telescopes that are now being developed will give us a more accurate idea of the size of the Universe in which we live.

The Earth as a planet

Together with Mercury, Venus, and Mars, the Earth—from the Greek word *eraze* that became the middle English word *erthe*—is one of the four planets closest to the Sun. These four planets are also called the terrestrial planets because they resemble the Earth in size (except for Pluto they are the four smallest planets in the Solar System) and also because they resemble the Earth in composition. All four are rocky, metallic planets.

Apollo 10 photograph of Earth in 1969.

The Earth has an equatorial diameter of 12,756 kilometers and a polar diameter of 12,714 kilometers. It can thus be thought of as slightly flattened or oblate in the polar diameter. This shape is the result of its rotation. It has a mass of 5.98×10^{24} kilograms, roughly the equivalent of 52,624,000,000,000,000,000,000 professional football linemen. The average density of the Earth is 5.5 times the density of water, making it the densest planet in the Solar System. Thus, most of the Earth must be fairly dense rocks and heavy metals such as iron and nickel.

In order to escape from the surface of the Earth, an object has to move 11.2 kilometers per second (or about 7 miles per second). For comparison: the fastest pitcher in the major leagues can throw a baseball only about 42 meters per second (about 95 miles per hour).

Depending on cloud cover, the Earth reflects 30 to 50 percent of the sunlight that strikes it. The moon reflects only about seven percent of the sunlight that hits it. The full Earth thus appears more than five times brighter as seen from the moon compared to the brightness of the full moon as seen from the Earth.

The temperature at the surface of the Earth varies considerably, from more than 100 degrees below zero Fahrenheit in the Antarctic to more than 135 degrees above zero Fahrenheit in the Sahara Desert. The overall mean temperature of the Earth is about 56 degrees Fahrenheit.

The structure of the Earth is such that it has a thin crust, only a few kilometers in thickness, composed of rock with a density 2.7 times the density of water. Below the crust is an extensive mantle, 29 hundred kilometers thick, composed of silicates whose density is estimated to average about 4.5 times the density of water. Closer to the center of the Earth is a liquid, outer core made of nickel and iron, on the order of 22 hundred kilometers thick and a density 9 times the density of water. Finally, there is a solid, inner core of nickel and iron that is thought to be 13 hundred kilometers in radius and whose density is about 12 times the density of water.

Twentieth century science has shown that the Earth's surface is made up of several massive plates that, on time scales of tens to hundreds of millions of years, slide around on the rocky but plastic boundary between the crust and the mantle. As these plates collide with each other and pull away or move by each other, earthquakes occur, mountains are pushed up, volcanoes emerge, and other geological activity results.

Because the Earth has an outer, liquid metallic core and because our planet rotates, a dynamo effect is created, giving our planet a substantial magnetic field. The Van Allen radiation belts, discovered by an Earth-orbiting satellite early in the spaceage, are a part of this magnetic field.

The atmosphere we have today is not the one we had 4.6 billion years ago when the Earth first formed early in the history of the Solar System. Our original atmosphere might have been an extensive envelope of hydrogen that quickly escaped into space. Outgassing from the Earth's interior then gave us an atmosphere dominated by carbon dioxide, methane, ammonia, and water vapor. The latter quickly condensed into the oceans. Over billions of years the atmosphere gradually evolved into the 80 percent nitrogen and 20 percent oxygen atmosphere we know today. Trace amounts of other gases, such as argon and carbon dioxide, are also present.

The Earth is the only planet in our Solar System that has been shown to have life. There is no direct evidence of planets orbiting stars beyond the Sun simply because these stars are too far away for our present telescopes to observe planetary systems. Therefore, as far as we know, our planet is the only one in the Universe on which life exists. There may be many other planets beyond the Solar System that have life, but we have no direct evidence of either the existence of these planets or, if they exist, of life on them.

The Earth in motion

The Earth rotates on its axis in a counterclockwise direction, as seen from north of the Solar System, with respect to the Sun once every 24 hours. Since the Earth also revolves around the Sun, the Earth rotates on its axis with respect to the stars in about 23 hours and 56 minutes. A sidereal day, based on the stars, is thus about four minutes shorter (three minutes and 56 seconds) than the solar day based on the Sun.

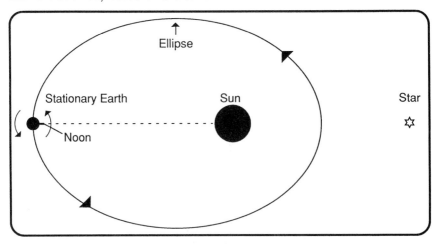

The Earth rotates on its axis once every 24 hours with respect to the Sun.

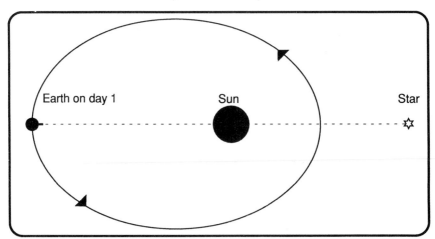

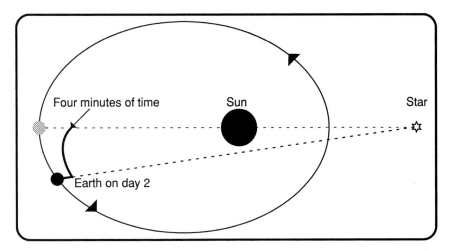

Four minutes of time   Sun   Star

Earth on day 2

The Earth rotates on its axis once every 23 hours and 56 minutes with respect to the stars.

There are about 365 days in a year, so we can calculate that three minutes and 56 seconds times 365 equals 1,436 minutes. This figure divided by 60 (the number of minutes in an hour) equals 24 hours, an entire day. Thus, at the end of a year we are back where we started. That is, there are about 365 solar days and 366 sidereal days in one year.

Thus, all the stars in the sky that rise over the eastern horizon do so about four minutes earlier each day. The result is that the stars and their constellations change with the passing seasons. For example, constellations we see in the evening sky in winter—Orion, Gemini, Canis Major and Canis Minor, and Taurus, for example—are not visible to us in the summer sky because they are above our horizon during the daytime. Conversely, the summer constellations—the Summer Triangle, Cygnus, Boötes, Lyra, and Aquila—are not visible in the winter because they are up during daytime.

Some familiar constellations around the northern part of the sky—the Big and Little dippers (the Big Bear and Little Bear), Draco, Cepheus—are circumpolar and thus never rise or set as seen from the mid-latitude of the Northern Hemisphere. We see them any night of the year, but at varying positions depending on the time of night and the season of the year.

At the Earth's equator, which is about 40 thousand kilometers in length (25 thousand miles), the speed of rotation is obviously 40,000 kilometers every 24 hours, or about 1,667 kilometers per hour (a little more than 1,000 miles per hour). At the precise North and South poles of the Earth, the speed of rotation is zero. Everywhere else on the Earth the speed of rotation is between zero and 1,667 kilometers per hour, depending on latitude.

Since the Earth's orbit is not a circle but rather an oval-shaped path, called an ellipse, our planet is at constantly varying distances from the Sun. By Kepler's second law of planetary motion, the Earth must also move at constantly varying speeds. The Earth is farthest from the Sun (aphelion) about the first of July and closest to the Sun (perihelion) about the first of January. Thus our planet's orbital speed is 30.3 kilometers per second at perihelion and 29.3 kilometers per second at aphelion, for an average orbital speed of 29.8 kilometers per second.

Due to the Earth's revolution around the Sun, the Sun seems to move from west to east against the background of stars.

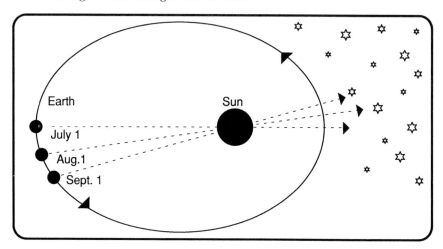

The Sun's apparent path against the background of stars is called the ecliptic.

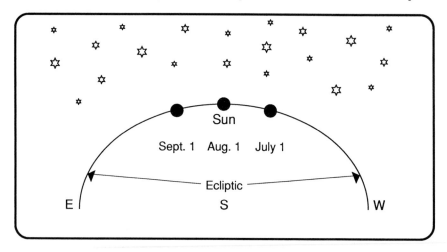

Obviously, the Sun requires a year to complete one trip on the ecliptic. The apparent motion of the Sun through the sky on the ecliptic is not proof of the Earth's revolutionary motion. If we lived in an Earth-centered (geocentric) Solar System, we would observe the same phenomenon.

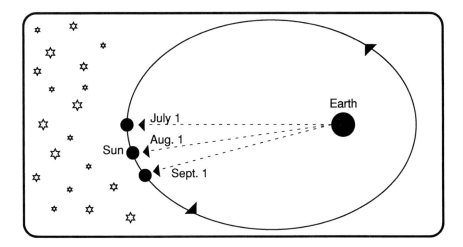

In addition to rotation and revolution, the Earth has a third major motion called precession. Think of a gyroscope spinning on its axis of rotation. As it does so, it also wobbles, or precesses. The Earth moves in exactly the same way.

It takes the Earth about 25,800 years to complete one wobble, or precessional cycle. If we were to extend the Earth's axis of rotation to the north, it would point very nearly at Polaris, the North Star. Due to precession, however, the star Vega in the constellation Lyra will be the North Star in 14,000 years.

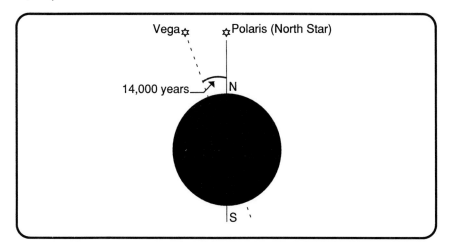

It is now thought that the periodic ice ages are the result of slight changes in the Earth's motions that we have just described. For example, the oval shape of our orbit varies over time, thus changing the amount of sunlight we receive. Further, our orbit itself revolves around the Sun, changing the times during the year when the seasons occur relative to the aphelion and perihelion points on our orbit. These two changes in our orbit are the result of the gravitational influences of the other planets in the Solar System.

Moreover, the tilt of our equator with respect to the plane of our orbit around the Sun (the ecliptic) also changes over time.

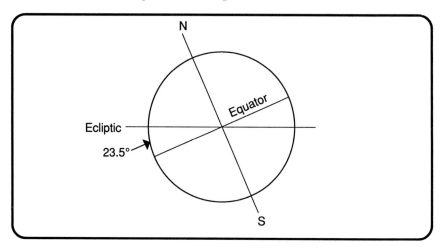

Currently this tilt is 23½ degrees, but this figure can vary by a degree or two either way. This variation means that higher latitudes receive more or less sunlight, depending on the angle of the tilt. Polar caps freeze farther into lower latitudes, helping to generate ice ages, or retreat into higher latitudes, helping to create interglacial periods. The Earth is in an interglacial period now, the last massive ice sheets in the Northern Hemisphere having disappeared about 10 thousand years ago.

Together, all of these changes in the Earth's orbit influence our planet's long-range climate conditions.

Easter

This Christian religious day seems to wander around the calendar with no set rules for its occurrence. In fact, the date of Easter is fixed by custom in Christian churches based on two astronomical events: the date of the vernal equinox and the occurrence of a full moon. The rule for establishing the date is that Easter always falls on the first Sunday after the first full moon that occurs on or just after the vernal equinox. The vernal equinox we know more familiarly as the first day of spring. It is that date on which the Sun, in its apparent annual trip through the sky, crosses the equator as it moves from south to north. Days become longer than nights, and our thoughts return to spring, growth, and the outdoors. The date of the first day of spring varies slightly, by a day or two, but generally occurs on March 21 each year. Because the full moon can occur on any one of many days after the vernal equinox, the date of Easter can be as early as March 22 and as late as April 25.

Phases of the moon, including the full moon, have no correlation with the vernal equinox. That is, the moon revolves around the Earth independent of the revolution of the Earth around the Sun. It is the latter motion that is responsible for the changes in the position of the Sun, and hence the occurrence of the vernal equinox, throughout the year.

Our calendar, of course, is not based on the phases of the moon but rather on the movement of the Sun: our Earth year is reckoned by the return of the Sun to the same place in the sky each year. Not all cultures have lived by a solar year. Some have tried to live by a lunar calendar, and there were many attempts in ancient times to reconcile the solar year with the phases of the moon. These attempts have always proved futile because the period of revolution of the moon about the Earth does not divide evenly into the period of revolution of the Earth about the Sun. In other words, the two periods of revolution are always out of phase with one another and cannot be numerically reconciled.

It is interesting to note, however, that we still have in our calendars a remnant of the attempt to correlate the movements of the Sun and moon: we recognize a division of the solar year into months—the words moon and month have the same origin. As long as we are going to have a solar calendar, however, there is no reason to divide it artificially into months. This practice is simply a historical custom with no modern practical meaning or necessity.

The word vernal means green, as in Vermont, the Green Mountain State. In spring, of course, the land becomes green again, a very important circumstance to people who depend on agriculture for their existence. The word equinox means equal night, referring to the two times of the year, spring and fall, when night and day are of equal length.

The origin of the word Easter is unclear, but it may derive from the Anglo-Saxon goddess of spring whose name was Eostre. In any event, Easter is not only an important Christian day of celebration signifying rebirth and re-surrection, but it is also a time for all people to celebrate the rebirth of life as the promise of warmth and vitality once again returns to the land.

Eclipses of the moon: Why and when

During a full moon, when the Earth is between the Sun and the moon, an eclipse of the moon is possible (see drawing p. 44).

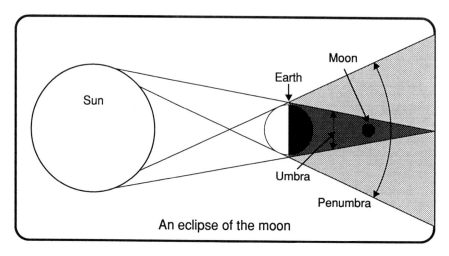

An eclipse of the moon

As the moon revolves around the Earth it first enters the penumbra of the Earth's shadow; the penumbra is the partially shaded portion of the shadow that surrounds the dark umbra. We cannot usually notice a difference in the light reflected from the full moon during the penumbral part of the eclipse because there is too much light in this part of the Earth's shadow.

When the moon enters the umbra, a curved shadow will begin to creep across the brightly lit full moon. The curved shadow on the moon's surface demonstrates the curvature of the Earth; this phenomenon was one of the proofs of the spherical shape of the Earth cited by Aristotle.

Totality is the condition when the moon is completely in the umbra of the Earth's shadow. It can last for a maximum of one hour and 40 minutes. The moon is never completely dark during totality because sunlight is refracted through the Earth's atmosphere onto the surface of the moon:

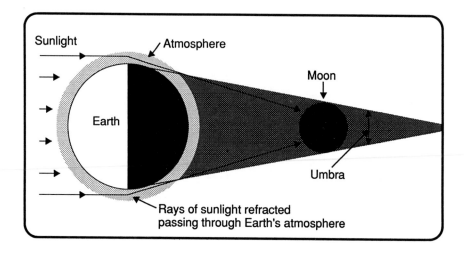

Light reflected from the surface of the moon can range from a dark reddish color to a lighter, coppery color; variation in color is due to weather conditions in the Earth's atmosphere where the sunlight passes through, as well as to the amount of particulate matter from volcanoes and pollution.

There are three kinds of lunar eclipses: penumbral eclipses occur when the moon passes through the penumbra of the Earth's shadow and completely misses the umbra; a partial eclipse occurs when the moon passes through only part of the umbra without going entirely into the dark part of the shadow; and a total eclipse, when the moon is completely inside the umbra.

Obviously, an eclipse of the moon can occur only during a full moon. Since there is a full moon each month, we would expect to see one of the three kinds of eclipses each month. However, we know that an eclipse of the moon does not occur each month. The reason is that the moon in its path around the Earth most frequently passes either above or below the Earth's shadow; only occasionally will the moon move directly into the Earth's shadow.

Eclipses of the Sun: Why and when

Eclipses of the Sun can occur only during a new moon, when the moon is between the Earth and the Sun:

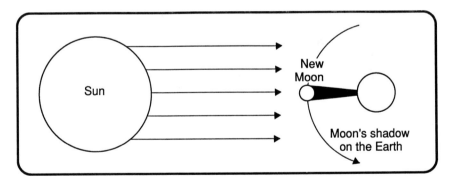

A total solar eclipse can only be seen from that part of the Earth where the cone of the moon's shadow falls on the Earth. The maximum width of the shadow is 269 kilometers. Since the moon revolves around the Earth at about 34 hundred kilometers per hour, the tip of the moon's shadow sweeps along the surface of the Earth at the same speed. Therefore, totality at any one spot is quite short, having a maximum duration of $7\frac{1}{2}$ minutes. A partial solar eclipse will be seen by people on either side of the moon's shadow cone.

An annular eclipse occurs when the moon's shadow does not quite reach the Earth's surface, a condition that occurs frequently (see drawing p. 46).

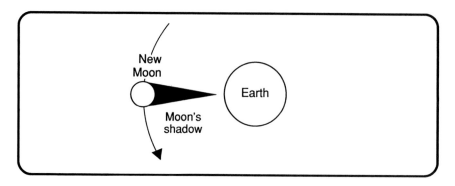

The word annular derives from the Latin word *annulus*, meaning ring. Because the moon will not quite cover the disk of the Sun under this condition, an annulus, or ring, of the edge of the Sun will be seen around the edge of the moon.

During totality, the sky is darkened, and stars are visible. Some flowers close, and chickens and other birds roost. The atmosphere and landscape take on eerie and strange colors.

Solar corona photographed during the total solar eclipse of August 31, 1932, at Freyburg, Maine.

Unlike lunar eclipses, which have little scientific value, solar eclipses are studied intensively by astronomers. The Sun's atmosphere, called the corona, is normally lost in the glare of sunlight; during a total solar eclipse, however, the corona is visible around the edge of the moon and can be studied directly. We can also learn about the light-scattering properties of our own atmosphere. In addition, we can check Einstein's theory of gravity by measuring the position of stars that appear close to the Sun; only during a total solar eclipse can stars be seen in the same part of the sky where the Sun is.

We do not have a solar eclipse each month during the new moon, even though there is a new moon each month, because the moon most frequently passes either above or below the Sun as seen from the Earth.

WARNING: Should you be fortunate enough to be in the path of a solar eclipse, never look at the Sun directly. Rays from the Sun can do permanent damage to your eyes. Always project an image of the Sun through a telescope onto a piece of white cardboard, or use a pinhole in a piece of cardboard for projection.

Einstein

Born in Switzerland in 1879, Albert Einstein has done more to shape our ideas about the nature of the Universe than any other scientist since Isaac Newton. In 1905, Einstein published what has since been called the Special Theory of Relativity. The implications of this theory are that time, length, and mass—basic concepts of the Universe—do not have absolute meaning as Newton firmly believed. Time, for example, depends on the relative motion of an observer to a clock in another frame of reference. The length of an object such as a ruler also depends on the relative speed of motion between an observer in one reference frame and an observer in a second reference frame.

The faster an object moves relative to an observer, the greater its mass becomes and the shorter it becomes in the direction of its motion relative to the observer. Out of this idea grew Einstein's famous equation, $E=mc^2$, that relates mass to energy by showing that they are two forms of the same thing. A small amount of mass, m, can be turned into a large amount of energy because m is multiplied by the square of a large number, c, the speed of light (three hundred thousand kilometers per second). It is this relationship that describes how stars shine: they convert mass into energy in their cores precisely as described by Einstein.

Relativistic effects—increase in mass, and time and space dilation—are noticeable only when the relative speed between an observer and an object or a clock approaches the speed of light. Newtonian physics, although not quite correct, works perfectly well in our everyday world where most of us do not deal with objects that travel close to the speed of light.

Einstein also showed us that space and time are bound together. This idea led to his next great insight about the Universe, the General Theory of Relativity. Published in 1916, it is the new theory of gravitation, replacing the older Newtonian theory that was based on force. Newton theorized that the force of attraction between any two masses—for example, the Earth and the Sun, or between the Earth and the moon, or between your body and the Earth—is directly proportional to the product of those two masses and inversely proportional to the square of the distance between them. We express Newton's law of gravitation as

$$\frac{\text{Force of attraction}}{\text{between any two masses}} = \frac{G \text{ (Mass One) (Mass Two)}}{\text{Distance}^2}$$

We have to put in G, the universal constant of gravitation, to balance the equation.

Einstein replaced the idea of force in gravitation theory with the idea that mass curves spacetime. The greater the mass, the greater the curvature of spacetime. The Earth, for example, is not held in orbit around the Sun by a gravitational force, as Newton believed. In Einstein's interpretation the Earth moves around the Sun in the natural curvature of spacetime created by the Sun's mass. The moon moves in the spacetime of the Earth's mass (i.e., the moon orbits the Earth), and you are held on the Earth not by a force of gravity but by the curvature of spacetime of the mass of the Earth.

All experiments conducted to date to test both the Special and General theories of relativity show that Einstein is correct. For example, in 1919, careful measurements were made of the positions of a group of stars, first in the nighttime sky and then again in the presence of the Sun during a total solar eclipse when the light from these same stars could be seen passing close to the Sun (a significant mass). The starlight was shifted as it passed through the spacetime near the Sun by the amount predicted by Einstein.

Einstein's theories are now used by astronomers to explore the large-scale characteristics of the Universe, to analyze black holes, neutron stars, the early Universe just after the Big Bang, and to make predictions about the future of the Universe.

This century's most prominent physicist—and one of the greatest scientists of all time—died in 1955. He left us a legacy of fundamental knowledge that will help us understand the structure of the Universe, its past history, and future evolution.

Foucault pendulum

Many readers have been to planetariums or museums that display a Foucault pendulum, a device that demonstrates—and is a direct proof of—the rotation of the Earth on its axis.

In 1851, a French scientist, Léon Foucault, suspended an iron ball weighing 25 kilograms (about 55 pounds) from the dome of the Pantheon in Paris at the end of a wire more than two hundred feet in length. The wire was attached to a device that allowed it to swing freely in any direction. He caused the iron pendulum to swing back and forth over a ring of smooth sand on a table. A stylus on the bottom of the pendulum cut a ridge through the sand every time the ball swung back and forth.

Within a few hours it was clear, because the stylus was cutting a series of ridges that progressed slowly around the sand, that the plane through which the pendulum was swinging was changing with respect to the sand, the table, and the Pantheon. Of course, it was not the plane of the pendulum's swing that was changing, but rather the Earth turning underneath the freely swinging pendulum. Foucault had thus proved that the Earth rotates on its axis, causing the table of sand to appear to turn under the pendulum.

This phenomenon is better understood if we imagine a pendulum freely swinging in its plane at the Earth's North Pole. As it swings back and forth, the Earth turns under it once in 24 hours, the period of rotation of the Earth.

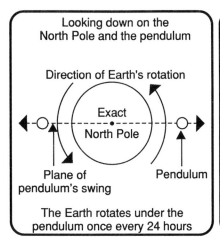

Looking down on the North Pole and the pendulum

Direction of Earth's rotation

Exact North Pole

Plane of pendulum's swing   Pendulum

The Earth rotates under the pendulum once every 24 hours

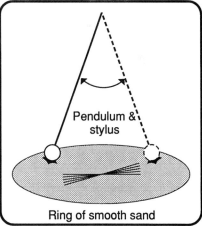

Pendulum & stylus

Ring of smooth sand

If the pendulum were to be put on the Earth's equator, there would be no motion under the pendulum. Thus, any point between the North Pole and the equator takes longer than 24 hours for the pendulum to appear to make one complete series of swings. This complete series of swings is called the period of the pendulum.

Galaxies: Physical characteristics

Galaxies are collections of stars, gas, and dust held together by gravity. They contain millions to hundreds of billions of stars. One nearby group of stars, the Andromeda Galaxy, contains an estimated three hundred billion solar masses. That is, the mass of this galaxy is about three hundred billion times the mass of our star, the Sun. Sizes of galaxies range from perhaps a few hundred light-years in diameter (one light-year equals about six million, million miles) to hundreds of thousands or even millions of light-years in diameter.

The closest galaxies to the Milky Way are the Large and Small Magellanic clouds; they are 163 thousand and 200 thousand light-years from our galaxy, respectively. Neither of these galaxies is visible from our part of the world, but they are quite prominent in the southern sky. The closest galaxy we can see from the Northern Hemisphere is the Andromeda Galaxy, visible to the unaided eye as a faint, fuzzy patch of light in the constellation of Andromeda. It can be seen only on a clear, dark night.

No one knows how far away the most distant galaxies are. Measuring enormous distances in the Universe is difficult for astronomers to do, and the accuracy of their measurements at this time is thought to be poor. It is estimated that the faintest, and presumably the most distant, galaxies are several billion light-years from the Milky Way.

The number of galaxies that can be photographed through the largest telescopes is estimated to be at least a billion objects, and the number may well be tens or hundreds of times this many galaxies within the Universe that we can observe. Many of these galaxies are smaller than our Milky Way; many are larger. We conclude that ours is an average-size galaxy.

It appears that nearly all galaxies are members of a cluster of galaxies. Regular clusters of galaxies contain at least one thousand members. Irregular, or open clusters, contain no central nucleus of galaxies; they appear to be simply a loose collection of galaxies without the symmetrical shape found in regular clusters. Open clusters generally contain fewer galaxies than regular ones.

The Local Group is an irregular cluster and contains at least 20 galaxies, dominated by the Andromeda Galaxy and the Milky Way, spread over a region three million to four million light-years in diameter.

There also exist clusters of clusters of galaxies, sometimes called superclusters. A local supercluster, including the Milky Way in the Local Group, has been identified. It has a diameter between 150 million and 300 million light-years. This local cluster of clusters of galaxies is thought to have a combined mass of 10^{15} solar masses. Some astronomers think that superclusters are the largest organizations of matter in the Universe.

Galaxies: Shapes

Astronomers have classified galaxies in the Universe by their shapes. The most common shape is the elliptical galaxy; these galaxies range from spherical to elongated egg shapes and contain mostly old stars. They have little dust, gas, and other matter from which new stars are formed. Most elliptical galaxies either do not rotate or rotate slowly. A few elliptical galaxies have masses that indicate that they contain one hundred times more stars than the Milky Way Galaxy. (Remember, the Milky Way contains an estimated four hundred billion stars.) Others are dwarf galaxies composed of only a few million stars.

Spiral galaxies have the best known galactic shapes. A typical spiral galaxy is characterized by one or more spiral arms made of stars, gas, and dust surrounding the central nucleus of stars. The arms contain young stars, and the nucleus contains older stars. About a third of the known spiral galaxies have bars through their nuclei; they are called barred spiral galaxies. The arms of these galaxies seem to originate at the ends of the bars.

In some spiral galaxies, including barred spirals, the arms are wrapped tightly around the nucleus; in others, the arms are much more loosely associated with the nucleus. Astronomers have further refined their classification of galaxies as a function of the mass (stars, dust, and gas) in the nucleus compared to the mass in the arms of galaxies. Spiral galaxies rotate so that the arms appear to trail behind the entire galaxy.

Our galaxy, the Milky Way, is a spiral galaxy, with the Sun and Solar System located in an arm of the Galaxy. It takes more than two hundred million years for our part of the arm of the Galaxy, including the Sun and Earth, to rotate once about the nucleus of the Milky Way.

There is also a small group of galaxies that are known as irregulars because they have no apparent shape. The Large and Small Magellanic clouds, the two closest galaxies to the Milky Way (visible in the Southern Hemisphere sky), are both irregular galaxies.

Many galaxies, particularly spirals, seem to have large halos of matter around them. This matter is not detected directly, but its indirect influence is observed in the rotational characteristics of galaxies. The nature of this matter is also unknown; it might be in the form of many small, dim stars too faint to be detected even by modern telescopes.

Some astronomers suggest that most galaxies have undergone, or are presently going through, significant interactions with other galaxies. They suggest that the observed galactic shapes are the results of galactic collisions and merger processes.

There are also a number of galaxies that appear to be ejecting matter from their nuclei. It is not yet clear to astronomers how this material is thrown out of a galaxy's nucleus.

Spiral galaxy in the constellation of Triangulum. Note that the arms are loosely associated with the nucleus of the galaxy. The nucleus is visible by light from old stars.

Located in the constellation of Ursa Major, this spectacular galaxy also has loosely associated arms wound around its nucleus. Light from hot, massive stars dominates the arms of the galaxy.

Lick Observatory Photograph

Also located in Ursa Major, this spiral galaxy has its arms more tightly wound around a larger, brighter nucleus. Our Milky Way Galaxy would appear similar to this head-on view.

Lick Observatory Photograph

Edge-on view of a spiral galaxy in the constellation Coma Berenices. The central nucleus is surrounded by arms that contain both stars and dust lanes. Our Milky Way resembles this galaxy from the same perspective.

Palomar Observatory Photograph

The Sombero Galaxy seen edge-on in the constellation of Virgo, as photographed by the 200-inch telescope at Palomar Observatory. Note prominent dust lanes in the arms of the galaxy.

Barred spiral galaxy seen head-on.

Interacting galaxies in the constellation of Corvus. These two galaxies may be colliding, thus rearranging the distribution of stars, gas, and dust.

Lick Observatory Photograph

The Whirlpool Galaxy in the constellation Canes Venatici. Most of the mass of the galaxy is contained in its arms.

Galileo

Galileo Galilei (1564-1642), an Italian and one of the greatest scientists of all times, was trained in medicine, mathematics, and astronomy. He was professor of mathematics at the University of Pisa and taught both mathematics and astronomy at the University of Padua. He became a follower of the Copernican system; that is, he gave up the idea that the Earth is the center of the Universe and defended the theory that it is one of several planets revolving around the Sun.

Although Galileo did not invent the telescope, he was one of the first scientists to use it in a systematic way to observe the moon, Sun, planets, Milky Way, and stars. One of his most important astronomical discoveries was the four inner moons of Jupiter. This significant discovery supported the Copernican concept of the Universe because it demonstrated that celestial objects revolve around other planets, much as the moon revolves around the Earth, and are not compelled to orbit the Earth as ancient astronomers and philosophers had taught.

Another of Galileo's discoveries that supported Copernican ideas about the Universe was the phases of Venus, similar to the phases of the moon. Galileo correctly attributed the planet's phases to its motion around the Sun, thus demonstrating that at least one planet goes around the Sun and not around the Earth as nearly all other people of Galileo's time believed.

With his telescope Galileo saw sunspots on the Sun, and mountains, valleys, and craters on the moon. By the motions of the sunspots, he showed that the Sun rotates on its axis. Although he was incorrect in interpreting the dark, flat plains of the moon as oceans (he named them *mare*, the Latin word for sea or ocean), he did suggest that the moon, with its mountains and valleys, resembles the Earth, thus categorizing the Earth as an astronomical object and denying it special creation. Again, Galileo dealt a blow to ancient ideas. With his telescope he also discovered that there are stars that cannot be seen with the unaided eye and that the Milky Way consists of a myriad of stars.

Because he was a radical thinker, challenging conventional interpretations of the Universe that had dominated people's thinking for 15 hundred years, he fell into disfavor with the established political and religious bureaucracies of his day. He stood trial for his astronomical ideas and interpretations that he based on observational and experimental evidence, was found guilty, and subsequently was condemned to house arrest for the last 10 years of his life.

Globular clusters

Globular clusters are aggregations of stars, bound together gravitationally, that form a shell around our galaxy and other galaxies whose details we can

Palomar Observatory Photograph

The globular star cluster in the constellation of Hercules. The light from hundreds of thousands of stars composes this image.

study. They are called globular clusters because they are spherical or globular in shape.

The smallest globular clusters contain several tens of thousands of stars, perhaps several hundred thousand stars. The largest globular clusters have a few million stars. All the stars are old, red dwarfs and red giants, indicating that the globular clusters themselves must be quite old. Indeed, the stars in the globular clusters surrounding the Milky Way are on the order of 15 billion years old, older than the Galaxy itself. We do not understand how the globular clusters formed or how they participated in the formation of the Galaxy.

There is little, if any, gas and dust associated with globular clusters. Most of this material must have been swept up long ago in star-forming processes; no stars appear to be in the making in globular clusters, thus supporting our own interpretation that these kinds of clusters are very old.

Globular clusters are also metal poor, again indicating great ages for their stars. (To an astronomer, any element other than hydrogen and helium is called a metal, not in the chemical sense but rather in the sense that hydrogen and helium make up about 99 percent of the mass of the Universe.) Any group of stars that is old must be metal poor because the stars were formed only out of primordial hydrogen and helium long before the heavier elements were synthesized through nuclear fusion in stellar cores.

Since globular clusters have diameters ranging between 30 and 60 light-years, the average distance between stars in a typical globular cluster can only be about one-third of a light-year. We remember that the distance between the Sun and its closest stellar neighbor is $4\frac{1}{3}$ light-years. Thus, were the Solar System in a globular cluster, the nighttime sky would be a spectacular blaze of starlight.

Many globular clusters are also the source of x-rays, probably from binary star systems in the center of these objects.

The Hercules globular cluster is a spectacular sight in our summer sky. (It is called by this name because it is in the constellation of Hercules.) Under dark, clear skies it can be seen by the unaided eye as a faint, fuzzy patch of light. Through even a small telescope it is a magnificent sight. This globular cluster is 21 thousand light-years distant.

There are about 160 globular clusters known to be associated with our galaxy. The number is probably larger, however, since we cannot observe many parts of the Galaxy due to intervening clouds of dust.

Gravitation: Newtonian style

As you are reading this book you are probably sitting in a chair. Were it not for gravitation, the tendency of matter to attract and to be attracted to other matter, you would be bouncing around the ceiling of the room.

Your body has mass; it is composed of electrons, protons, and neutrons that in turn are the building blocks of the many kinds of atoms that make up the molecules of your hair, skin, bones, organs, blood, fat, and muscle. Your mass is attracted to the center of the Earth's mass. Gravitation is also a mutual force: the Earth's mass is attracted to your mass. Because the Earth has much more mass than you do, however, it is you who are held to the surface of the Earth rather than the Earth being attracted to you with any noticeable effect.

Isaac Newton was the first scientist to give definition to the concept of universal gravitation. He reasoned that the same force that causes an apple to fall from a tree to the Earth also holds the moon in orbit around the Earth and the Earth in orbit around the Sun.

Quite simply, he suggested that every particle, every mass in the Universe, attracts every other mass. Your body and the Earth are not only attracted toward each other, but the mass of your body is also attracting—and being attracted by—the moon, Sun, all the stars, and the most distant galaxies.

To understand why you are not flying through space to the moon, Sun, or some distant galaxy, we have to examine Newton's description of gravitation in more detail. Newton reasoned that the force of attraction between two masses could be stated as the product of the two masses divided by the square of the distance between them. The force of attraction decreases as the distance between the masses increases. Even though the mass of the Sun is much larger than the mass of the Earth, you are much closer to the mass of the Earth than you are to the mass of the Sun and are thus more strongly attracted to the Earth.

The mass of the Andromeda Galaxy is perhaps three hundred billion times more than the mass of the Sun; yet, we are not shooting through space to this galaxy because its distance from us is so great (more than two million light-years). In the same way, you weigh slightly less at the top of the Empire State Building in New York City than you do on the first floor of the same building because your mass and the Earth's mass are farther apart when you are on top of the building.

People are sometimes confused about the terms weight and mass. Weight is a measure of force due to gravitational attraction. For example, if you weigh 180 pounds on the Earth, this weight is a measure of the force of gravitational attraction between your mass and the mass of the Earth. If you go to the moon with your scales, you will find that they will register only about one-sixth the weight (about 30 pounds) they did when you stepped on them on the Earth.

Weight watchers take note! When you go to the moon you still have as much mass in the form of fat, bones, blood, and other material as you had on the Earth. The difference in weight is due to the difference in mass of the Earth and the mass of the moon. The moon, having less mass than the Earth, consequently causes the force of attraction between you and the moon to be smaller than between you and the Earth, thus depressing the scales less and showing less weight for your mass. Going to the moon will only help you lose weight, not mass.

Gravitational lenses

Quasars are the most distant and also the brightest objects that we know about in the Universe. They are almost certainly the active nuclei of massive galaxies. We observe only the very bright nuclei of these galaxies because quasars are so far away that the rest of their galactic structure is too dim to be seen. In the billions of years it takes their light to reach us here on the Earth, it is probable that this light would encounter other massive objects such as galaxies or clusters of galaxies that are closer to us.

Einstein's General Theory of Relativity tells us that in the presence of a large mass, spacetime is noticeably curved or warped. It would then seem logical that light from a distant quasar passing by a significant mass, a galaxy or a cluster of galaxies, would be bent by the curved spacetime of this mass.

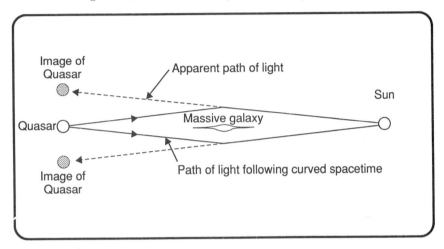

As a result, we observe a double or multiple image of the quasar. The gravitational field of the galaxy or cluster of galaxies acts like a lens to produce more than one image of a single object.

How do we know it is the same quasar and not just the happenstance of two quasars close to each other in the sky? The answer is that if we find two quasars close to each other in the sky that have identical spectra, including identical redshifts that indicate the same recessional velocities, then we are justified in concluding that we are seeing two images of the same quasar. The intervening galaxy or cluster of galaxies that causes this phenomenon need not even be visible to us; generally, only its effects (the double or multiple images) are observed.

The first gravitationally lensed quasar was discovered in 1978 with the Multiple-Mirror Telescope on Mt. Hopkins in Arizona. There are about a dozen gravitational lensing systems that have been discovered to date.

GUTS (Grand Unified Theories)

There are four known forces that account for the behavior and activity of all matter and energy in the Universe. We are familiar with the force of gravity in our everyday lives. Gravity operates over infinite distances, but is the weakest of the four forces. The reason that gravity is so noticeable is that it depends on mass, and there is a lot of mass in the Universe.

The strongest force in the Universe is the strong nuclear force, some 10^{39} times stronger than gravity. It is responsible for holding together the protons and neutrons in the nucleus of the atom. The strong nuclear force operates over very short distances, on the order of $1/10,000,000,000,000$ of a centimeter (10^{-13} centimeter). Because it works only on the subatomic level, we do not notice it in our daily activities.

The other force that is effective only on the subatomic level is called the weak nuclear force. It operates over even shorter distances, $1/1,000,000,000,000,000$ of a centimeter (10^{-15} centimeter). It is one hundred thousand times weaker than the strong nuclear force, but still 10^{34} times stronger than gravity. The weak nuclear force is responsible for radioactivity and for fission processes in atoms.

The fourth force is called the electromagnetic force; it keeps atoms together by binding electrons to protons. The electromagnetic force is responsible for all the radiative energy in the Universe: visible light, infrared, ultraviolet, x-ray, gamma ray, radio, television, and microwave radiations. We are thus quite aware of this force in our lives.

The electromagnetic force resembles the gravitational force in that it operates over infinite distances, but it is 10^{36} times stronger than gravitation. Gravity, as we know, always attracts, whereas the electromagnetic force can either attract or repel. That is, alike electrical charges repel each other, and opposite electrical charges attract each other.

Einstein worked unsuccessfully for years to find a satisfactory way to bring together, physically and mathematically, his theory of gravitation with the other forces in the Universe into a unified field theory. It is just in the past several years that theoreticians have been successful in showing that, in the very early Universe, at least three of the forces were unified.

If we imagine the Universe at 10^{-43} of a second after the Big Bang (this expression of time in the very early Universe can be written as $1/10,000,000,000,000,000,000,000,000,000,000,000,000,000,000$ of a second), all four forces were united into what some scientists call the superforce. In other words, the four forces were indistinguishable from each other. The temperature at this early epoch was on the order of 10^{32} ($100,000,000,000,000,000,000,000,000,000,000$) degrees Kelvin. There was no matter in the Universe, only radiation.

As the Universe expanded due to the Big Bang, it also cooled. The various forces "froze" out and became recognizable as the ones we observe today. The first force to freeze out was gravity at 10^{-43} of a second. When the Universe was 10^{-35} of a second old and had a temperature of only 10^{27} degrees Kelvin, the strong force froze out, leaving the electroweak force, a combination of the electromagnetic force and the weak nuclear force. As the Universe continued to expand, the temperature dropped to 10^{16} degrees Kelvin at 10^{-12} of a second when the electromagnetic and weak nuclear forces froze out. At this period, then, all four forces that we recognize today were separate and distinguishable.

Theorists are confident that the strong, weak, and electromagnetic forces were unified at 10^{-43} of a second into the history of the Universe. This circumstance is described by several Grand Unified Theories (GUTS). No one has yet been able to show that at an earlier epoch gravity was also a part of the unification, although some of the best scientific minds are working on the problem. It is amazing, however, that modern physical theory and observation have given us a sound understanding of the early Universe back to 10^{-43} of a second from the Big Bang.

Halos of the Sun and moon

Occasionally, when we look into the daytime sky in the vicinity of the Sun or in the nighttime sky near the full or nearly full moon, we see a faint halo of light around these two objects. This faint ring of light appears to have a radius of about 22 degrees; that is, the angular distance from the Sun or moon to the ring is 22 degrees as we see them from our position on the surface of the Earth. There may also be an outer ring—quite faint and hard to detect—that has an angular distance of 46 degrees from the Sun. (The moon shines too faintly for us to observe a second halo.)

The inner edge of the halo is always rather sharply defined and has a faint, red tint. Succeeding colors from the inner to the outer portion of the halo are, in order: faint tints of orange, yellow, green, blue, indigo, and violet. The outer edge of the halo is fuzzy and diffuse; colors in this part of the halo are difficult to detect.

The cause of the solar or lunar halos is ice crystals in cirrus clouds in the Earth's atmosphere at an altitude of seven kilometers (about four miles) or more above the surface of our planet. Light from the Sun or moon is bent (refracted) as it passes through these crystals; violet and blue wavelengths of light are refracted the most, and red wavelengths are refracted the least. Hence, from our position on the Earth we have the impression of a faint rainbow that completely circles the Sun or moon. No special arrangement of ice crystals is necessary to produce the halo.

Old-time observers suggested that halos around the Sun or moon, which of course can be seen only when the sky is clear, indicate a change in weather. There is some measure of truth in these predictions because high-altitude cirrus clouds often presage other kinds of clouds and perhaps even precipitation.

William Herschel

One of the best observational astronomers of all time, William Herschel, was born in Germany in 1738, but spent most of his life in England. He was a fine musician, but became interested in the stars early in his life. He was a self-taught astronomer, learning not only the astronomy and mathematics of the day, but also how to build his own telescopes. His largest telescope had a speculum metal mirror 4 feet in diameter with a focal length of 40 feet.

Using his own instruments, he analyzed the rotation of several planets and their moons. He was the first scientist to suggest that the polar caps on Mars were made of water ice, an observation confirmed in the twentieth century by space satellites. Herschel also determined the tilt of Mars's axis of rotation.

In 1781, he discovered the planet Uranus. As a result of this discovery, the King of England, George III, made Herschel his private, court astronomer, freeing Herschel to devote himself to astronomical research for the rest of his life. In 1789, Herschel discovered Uranus's two brightest moons, Titania and Oberon.

Through his telescopes Herschel found and catalogued more than eight hundred double stars, many of which turned out to be true binary systems: two stars revolving around their common center of mass and following Newton's universal law of gravitation.

Herschel also tried to determine the position of the Sun and Solar System in the Milky Way by doing star counts. He found that there were about the same number of stars wherever he looked along the Milky Way, and thus logically suggested that the Sun and Solar System are at the center of our galaxy. Today we know that this deduction is incorrect; there are clouds of dust in the Milky Way that obscure the light from large numbers of stars that would give us our true position far from the center of the Milky Way. It was not until the early part of the twentieth century that the Sun and Solar System were found to be in an arm of the Milky Way, about 28 thousand light-years from the center of the Galaxy.

Herschel also discovered infrared radiation, invisible to the human eye. Again, we know today that most objects in the Universe produce large amounts of infrared radiation, giving us information about their origin and evolution.

William and his son John catalogued more than five thousand nonstellar objects. We now know that these objects are external galaxies, planetary nebulae (exploding stars), clouds of hydrogen gas in our galaxy, and similar masses.

We should also mention Herschel's sister, Caroline (1750-1848), one of the world's first female scientists. She was also an outstanding astronomer who made many significant contributions to our understanding of the Universe. She reduced many of the data that came from William's observations, as well as data from her own observations. Data reduction in the days before calculators and computers often involved long, difficult, and tedious calculations by hand. Using her own telescope, she discovered eight new comets between 1786 and 1797.

For their superb scientific work, William was knighted and Caroline received the gold medal of the Royal Astronomical Society in 1828 and the gold medal for science from the King of Prussia in 1846 when she was 96 years old.

Hydrogen

Hydrogen is the most common element in the Universe. Present estimates suggest that 80 percent of the matter that we can see or otherwise detect in the Universe is hydrogen.

Hydrogen is also the lightest element, made up of one proton in the nucleus orbited by one electron. Because it has just one electron, hydrogen

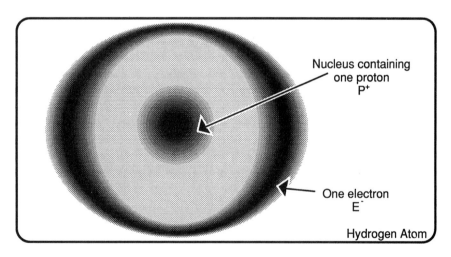

Nucleus containing one proton P^+

One electron E^-

Hydrogen Atom

readily combines, sometimes explosively, with many other elements. If two hydrogen atoms combine with oxygen, we have H_2O, water. If hydrogen combines with carbon, we have many hydrocarbon molecules, including CH_4, methane, or natural gas. If combined with nitrogen, we have NH_3, or household ammonia.

Most of the stars we see in the nighttime sky are made up of about 80 percent hydrogen, 20 percent helium, and trace amounts of the other 90 naturally occurring elements. Our star, the Sun, is about 80 percent hydrogen, 20 percent helium, and about one percent of all the other 62 elements that have been discovered there. Jupiter and Saturn are also mostly hydrogen; Neptune and Uranus have a great deal of hydrogen associated with them, although they also have other elements in abundance.

Thus, the Earth and the other terrestrial planets, Mercury, Venus, Mars, are anomalies in the Solar System and in the Universe because they have so little hydrogen. True, the Earth has a great deal of hydrogen in the oceans' water (H_2O), but this hydrogen is a tiny fraction of the mass of our planet, most of which is iron, silicon, and other heavy elements.

Nearly all the stars in the sky, including the Sun, shine because they fuse hydrogen nuclei into helium nuclei in their cores. The radiation that is released, when mass is converted to energy by $E=mc^2$, makes its way out of a star's core. Much of this radiation emerges at the star's surface as visible light—and the star shines in our sky.

Where did all the hydrogen come from in the Universe? The answer is that in the early Universe, as it expanded and cooled, protons and electrons formed out of hot radiation. As the Universe continued to expand and cool, electrons were able to join protons, and hydrogen was formed. Conditions were such that very few atoms of heavier elements were able to be made. Only helium and trace amounts of lithium were also found in the early Universe. Thus, hydrogen was, and remains today, the most common element in the Universe.

Since hydrogen is so prevalent, we expect that radiation given off by hydrogen might be common, too. Indeed, the most common radiation in the Universe has a wavelength of 21 centimeters, in the radio end of the spectrum. It originates from a hydrogen atom when its electron does a flip-flop and spins in the other direction relative to the spin of the proton in the nucleus. This energy is then released and observed here on Earth by radio telescopes.

Interferometry

A technique has been developed during the past three decades that makes radio telescopes powerful tools for probing the details of distant objects in the Universe. Radio telescopes have notoriously poor resolving power, or the ability to separate two points in the sky. Resolving power depends on both the size of the radio antenna (or "dish," which is equivalent to the lens of an optical telescope) and on the wavelength of the radio energy under study. The larger the antenna, or mirror, the greater the resolving power. Radio wavelengths are on the order of one hundred thousand times longer than optical wavelengths.

Astronomers have devised a way to make relatively small radio telescopes operate like large ones, by a process known as interferometry. If we take two small radio telescopes and separate them by a distance—say several kilometers—that is very large compared to their diameters, and arrange things so that they can both observe the same object in radio wavelengths at the same time, the two small radio telescopes will operate as two sides of a very large radio telescope. If radio waves arrive at the two antennas at the same time, they will be reinforced. As the object moves across the sky, due to the rotation of the Earth, the waves will arrive at the antennas at slightly different times, and a weaker signal will be the result. In other words, the waves destructively interfere with each other, and we have interferometry.

The two telescopes see the source object in the sky as a series of constructive and destructive signals—a pattern of reinforcements and inter-ferences. In this way, an image of the radio source can be built up.

Obviously, the wider the separation of the radio telescopes, the greater the resolving power. Radio telescopes on opposite sides of the Earth produce a resolving power of 1/1,000 of an arc second (1/3,600,000 of a degree) at 21 centimeters, which is equivalent to measuring the angle made by a quarter in Seattle as seen face-on from Miami. Thus, fine details of very distant objects that emit radio energy, such as radio-loud quasars and galaxies, can be observed.

The interstellar medium

Although the density is very low, the space between the stars is not empty. The interstellar medium, or ISM as it is referred to in the technical literature, in general is at a density much lower than our best laboratory vacuums. Yet, this diffuse material plays a big part in the formation of new stars, influences the way light travels across space, and provides us with some of the most spectacular sights in the Universe.

The density of the densest ISM is on the order of 10 thousand to 100 thousand (10^4 to 10^5) particles per cubic centimeter, compared, for example, to the density of the Earth's atmosphere at sea level, which is 10^{19} particles per cubic centimeter. (A sugar cube has a volume of about one cubic centimeter.) The temperature of these interstellar clouds is 20 to 50 degrees Kelvin (423 to 370 degrees below zero Fahrenheit)—very cold indeed.

Much of the ISM is less dense, containing only one to one thousand particles per cubic centimeter, with a slightly warmer temperature of 50 to 150 degrees Kelvin. The least dense parts of the ISM—in the majority of the space between the stars—contain only one particle in every 10 thousand cubic centimeters. The temperature of this very diffuse matter, however, ranges from one hundred thousand to one million degrees Kelvin. This high temperature is thought to be the result of shock waves from supernovae explosions.

The ISM is made up of about 90 percent gas and 10 percent dust. The dust is probably tiny particles of silicates and graphite (carbon), both of which may be covered with frozen gases including water vapor. The grains may also be dirty ice. It is thought that this dust forms in outer parts of the extended

A birthplace for thousands of stars, the Orion Nebula is one of the most spectacular and easily observed nebulae. The bright central region is the emission part of the nebula and the outer fringes are the reflection part of the nebula. The interspersed dark areas are dominated by dust that blocks and scatters starlight.

The Trifid Nebula in the constellation of Sagittarius. The bright emission nebula is divided by dust lanes.

Note the small, dark globules scattered in the bright emission part of the Lagoon Nebula in the constellation of Sagittarius. These dust and gas globules will probably collapse to form new stars during the next few million years.

The Horsehead Nebula in Orion is a mass of dust and gas with a temperature of only a few degrees above absolute zero. It is surrounded by gas with a temperature of many thousands of degrees. The nose of the horse may be collapsed by this hotter gas to form a new star.

A new star may be forming at the top of the Cone Nebula in the constellation of Monoceros. Palomar Observatory photograph by the 200-inch telescope.

Palomar Observatory Photograph

atmospheres of old stars where temperatures and pressures are just right to condense these materials.

The shorter wavelengths of starlight are scattered passing through interstellar dust. The result is that stars appear both dimmer and redder than they really are. These conditions must be taken into consideration when determining distances to the stars. Infrared and radio waves pass through the dust unimpeded and thus make good wavelengths at which to study objects in and beyond the dusty parts of the interstellar medium.

The gas part of the ISM can be in the form of emission nebulae. The radiative energy, particularly ultraviolet light, from nearby hot stars causes the gas to give off its own light, thus creating beautiful images. Sometimes light from hot stars is reflected off the gas and dust in the ISM, and we have a reflection nebula. There can be so much dust in the ISM that little visible light can pass through, and we have a dark nebula.

Gas in the ISM is made up mostly of hydrogen, either atomic or molecular. If this gas is giving off its own light, as in an emission nebula, we have what is called an H II region; that is, the hydrogen atoms are missing their electrons and are thus ionized. The second most common molecule in the gas part of the ISM is carbon monoxide. Methane, cyanide, water, sulfur dioxide, ammonia, formaldehyde, formic acid, methanol, and ethanol are also among the more than one hundred species of molecules found in the interstellar medium.

When these clouds of gas and dust of the ISM collapse, stars are formed. The hydrogen and other elements are processed into heavier elements by nuclear fusion in the cores of these stars.

When stars reach the end of their lifetimes, stellar winds or explosions return the processed material back to the ISM where it is again available for the formation of new stars. This means that the ISM is gradually—over billions of years—becoming heavier as it is composed of increasingly more massive atoms.

The inverse square law

If we record the brightness of a light bulb from a distance of one foot and then compare the brightness of the same bulb from a distance of two feet, we will discover that the brightness of the bulb at two feet is only one-fourth its brightness at one foot. Similarly, if we measure the brightness of the bulb at three feet, we will discover that it is only one-ninth as bright as it is at one foot.

In other words, we have discovered a simple law of nature: brightness falls off by the square of the distance as the distance increases. (The square of two is four [2 X 2 = 4], the square of three is nine [3 X 3 = 9], and so forth.) This law is called the inverse square law. This relationship is inverse because as the distance increases, the brightness decreases (there is an inverse, or opposite, relationship between brightness and distance). The relationship between brightness and distance is based on the square of the distance, not on some other mathematical ratio such as the cube or 1.5.

We can use this same principle in determining distances to stars. Suppose we have two stars that we know have the same intrinsic brightness (we know they are putting out the same amount of light because their spectra are identical), but one star is four times fainter than the other star. We thus know that the fainter star is twice as far away as the brighter star. Further, if we know the distance to the brighter star, we immediately know the distance to the fainter star.

We can apply the inverse square law to other objects to help us determine their distances. For example, suppose we have two galaxies that have nearly the same shapes, but one is nine times fainter than the other. At least as a first approximation, we assume that the intrinsic brightness of the two galaxies must be similar and that the fainter one must be three times farther away than the brighter one. If we know, through some other means, that the distance to the brighter one is 30 million light-years, we immediately know that the distance to the fainter galaxy must be about 90 million light-years.

The inverse square is apparent in another important physical relationship. In Newton's version of gravitation, the force of attraction between any two masses is proportional to the product of those masses divided by the square of the distance between them:

$$\text{Force of Gravitation} \propto \frac{(\text{Mass one})(\text{Mass two})}{\text{Distance}^2}$$

If the masses are large, the gravitational force is large; this is a direct relationship. If the distance between two masses increases, the gravitational force between them diminishes as the square of the distance; this is an inverse (or indirect) relationship. We can thus apply the inverse square law to help us calculate the masses of objects such as planets, moons, and binary stars, and to calculate the gravitational forces between them.

Thus it is that astronomers use the inverse square law in two significant physical relationships in their study of the Universe: the change in the force of gravitation and the change in the brightness of light, both as functions of the change of distance.

Jupiter

The fifth planet from the Sun, at a distance of 778 million kilometers, is Jupiter, the largest planet in the Solar System. The name comes from the chief Roman god, although the Romans had no way of knowing that Jupiter was the biggest planet. It has a diameter of about 143 thousand kilometers, slightly more than 11 times the diameter of the Earth. Its mass is about 318 times the mass of the Earth. Indeed, Jupiter is more massive than all the other objects, exclusive of the Sun, in the Solar System combined. (Jupiter has about 1/1,000 the mass of the Sun.) Its density is 1.3 times the density of water.

This Voyager 1 image of Jupiter from 37 million kilometers reveals details of the cloud belts and zones in the upper atmosphere of the largest planet in the Solar System.

The planet rotates on its axis once every nine hours and 55 minutes, much faster than the Earth (24 hours), so the average length of a day and a night on the planet is about five hours each. It takes nearly 12 Earth years to revolve once around the Sun. The giant of the Solar System planets also has more moons than any other planet (16), several of which can be seen with even a small, low-power telescope.

If we could stand on a scale on Jupiter, we would discover that we would weigh more than $2\frac{1}{2}$ times our weight on Earth. For example, if we weigh 150 pounds on the Earth, we would weigh more than 400 pounds on Jupiter.

Jupiter is made of mostly gases, the most common one being hydrogen. In addition to hydrogen, gases present are ammonia, methane, helium, and several others. Because Jupiter is so massive, it has a strong gravitational field, causing the hydrogen gas in its interior to be compressed so much that it becomes liquefied. Farther down in the planet, the hydrogen is so squeezed together that it may resemble a metal, or metallic hydrogen. There may also be a rocky core, perhaps several times the mass of the Earth, at the center of the planet. The lowest temperature at the tops of the clouds in Jupiter's atmosphere has been measured at 110 degrees Kelvin (about 261 degrees below zero Fahrenheit).

One interesting feature of Jupiter is that it radiates more energy at certain long wavelengths than it receives from the Sun. If we define a planet as an object that we see because it reflects sunlight, then Jupiter is a planet because all the visible wavelengths we see are reflected sunlight. If we define a star as an object that radiates its own light, then we would have to say that Jupiter has some star-like characteristics. This fact, combined with the knowledge that the relative abundances of the elements in the Sun and in Jupiter are similar, suggests that if Jupiter were about one hundred times more massive than it is, it could be considered to be a low-mass star.

Another interesting feature of the planet is the Great Red Spot, a large red area that changes size, shape, and intensity of color over time. The spot can have a maximum diameter of 50 thousand kilometers. It may be thought of as a storm in the upper part of Jupiter's atmosphere.

Jupiter also has a magnetic field that is more than 10 times the strength of the Earth's magnetic field. Charged particles such as electrons and the nuclei of atoms apparently whirl around these lines of magnetic force, not unlike the Van Allen radiation belt of the Earth, sending out energy at radio wavelengths.

Jupiter also has a thin ring system, so thin and tenuous that it is not visible through present Earth-based telescopes. It was discovered by the Pioneer 10 satellite.

Johannes Kepler

A number of scientists in Europe became Copernicans—they believed that the Sun is the center of the Solar System—soon after Copernicus died in 1543. Johannes Kepler (1571-1630) was a scientist and mathematician who, early in his life, became a convert to this heliocentric (literally, Sun-centered) system.

Kepler was a German, having been born in Württemberg, but spent much of his life in Prague as mathematician to Emperor Rudolph. While in this position, Kepler used the measurements calculated by his predecessor, Tycho Brahe, of the movements of the planets and of the positions of the fixed stars to work out three laws of planetary motion. (Tycho's measurements were quite accurate, considering they were made with various kinds of sighting instruments in days before the invention of the astronomical telescope by Galileo in 1609.)

Kepler's three laws of the motions of the planets can be summarized as follows:

1. The planets do not move in perfect circles, as many people believed, but revolve around the Sun in elongated orbits called ellipses. Thus, the planets are closer to the Sun at some point in their orbits than at others. The Earth, for example, is closest to the Sun in January and farthest away in July by a difference of more than five million kilometers.

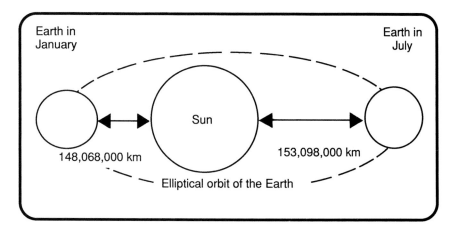

2. The closer a planet is to the Sun, the faster it moves; conversely, the farther a planet is from the Sun, the more slowly it moves in its orbit. In our diagram the Earth is moving faster in its orbit each day when it is closest to the Sun in January and slower when it is farthest from the Sun in July.

3. There is a relationship between the periods it takes any two planets to revolve around the Sun and their distances from the Sun. For example, if we know by observation the time it takes the Earth and Mars to revolve around the Sun and the distance from the Earth to the Sun, we can calculate the distance from Mars to the Sun. Kepler was the first astronomer to calculate this distance.

With these three laws, Kepler established a systematic, predictable relationship between the Sun and planets and, thus, helped lay the foundations for modern astronomy.

Life in the Universe

There are many definitions of life: that which reproduces itself, that which synthesizes protein from amino acids to create larger and more complex molecules, or that which metabolizes energy from the Sun or some other star. The list of definitions of life could be made much longer. Regardless of how we define life, however, we can say that life is a condition of the Universe, not only on our planet but also on a very large number of planets. Life exists throughout the Universe.

Having made this bold statement, we must hasten to add that the planet Earth is the only planet for which we have direct evidence that life exists. Indeed, the nine planets in the Solar System are the only planets we know of anywhere in the Universe by direct observation. The reason that we have not found any other planets is that even the closest stars are too far away for us to make direct observations with present telescopes of any planets these stars may have.

Furthermore, there is no evidence at the present time that life of any kind exists on the other eight planets in our Solar System. A few scientists have speculated that primitive forms of life (for example, subcellular life such as bacteria) may exist in the upper atmospheres of Jupiter or Venus, but these are merely speculations with no concrete data.

We have also sent two unmanned spacecraft, Vikings 1 and 2, to Mars to search for life. Neither spacecraft detected any firm evidence of life. Mercury is probably too close to the Sun for life to have evolved, and the other planets in our Solar System—Saturn, Uranus, Neptune, and Pluto—are too far away from the Sun for life to exist.

Therefore, our statement that life exists elsewhere in the Universe, beyond the Solar System, is based on inference, an indirect conclusion that considers the vast number of stars and, by implication, the very large numbers of planets that probably exist around these stars. No one knows how many stars there are in the Universe, but a number that is sometimes used suggests that there may be more than 10^{20} stars: that is a one with 20 zeros behind it. If only a tiny fraction of these stars have planets, and if only a tiny fraction of these other solar systems have the proper conditions for life to occur—for example, stars that exist long enough to allow life to evolve, stars that are stable enough to prevent life from being either fried or frozen, planets that are the right distance from their star so that they are within a range of appropriate temperatures for life to flourish, planets that are large enough to hold suitable atmospheres, and a wide range of other conditions that may be necessary for life—there would still be an enormous number, perhaps trillions and trillions of planets, on which life could exist.

Our argument for the existence of life elsewhere in the Universe is a numbers argument: it would be numerically next to impossible—a statistical improbability—that the planet Earth should be the only planet in the entire Universe on which life developed.

Whether or not this strong numerical evidence for the existence of life elsewhere in the Universe might also be used to suggest that intelligence has evolved elsewhere, however we define and understand intelligent life, is a question that is beyond the province of the astronomer. Indeed, this question may be beyond science in general to attempt to answer, at least with our current understanding of life and intelligence. Perhaps, these questions are best left to philosophers and humanists.

Light

All radiative energy in the Universe can be thought of as having various wavelengths. The shortest wavelengths are the most energetic and are called gamma rays. They have wavelengths of about 1/1,000,000,000,000 of a centimeter. Gamma rays are produced in the interiors of stars. Next are x-rays, with slightly longer wavelengths (about 1/1,000,000,000 of a centimeter), and ultraviolet light, with even longer wavelengths.

Visible light has wavelengths from 4/100,000 to 7/100,000 of a centimeter in length; the shorter wavelengths register on our eyes as blue light, and the longer wavelengths are red light. In between the blue and the red wavelengths are green, yellow, and orange, together with all the other colors that our eyes are capable of seeing.

Beyond the visible portion of this spectrum of energy, in increasing wavelengths (and hence decreasing energy), are infrared waves, microwaves, television waves, and finally radio waves. The longest radio waves can be several kilometers in length.

The speed of light, all the radiative energies we have been discussing, is both a constant and the limiting speed in the Universe: nothing can travel faster than light. The speed of light is designated by the small letter "c" and is related to wavelength by the simple equation: c = wavelength X frequency. Obviously, the longer the wavelength, the smaller the frequency (since c is a constant); and the shorter the wavelength, the greater the frequency.

We can visualize this situation in the following drawing:

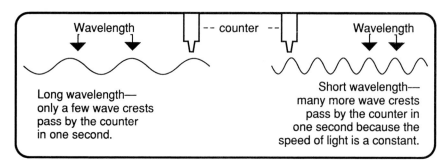

Light can be regarded as not only waves, but also as little particles of energy called quanta, or photons. If we measure light using a photometer (literally, a photon measurer), we are observing the photon, or particle, characteristic of light. Most of us are familiar with the photometers that are built into 35-millimeter cameras and give us exposure settings and times.

When light passes through a lens, as in a camera, telescope, eyeglasses, or our eyes, we are using the wave characteristic of light to refract (bend) light and bring it to a focus. When we divide the light of a star into its spectrum in a spectrograph attached to a telescope, we are also using the wave characteristic of light to analyze a star or a galaxy. Thus, we say that light exhibits a wave-particle duality. It is neither wave nor particle, but rather a combination of both wave and particle.

Astronomers explore all parts of the radiative energy spectrum for information about the many kinds of objects in the Universe. For example, radio telescopes have been developed in the past 40 years to search for data in the very long wavelengths of radio energy. Many stars (including the Sun) and galaxies are sources of radio energy. By studying the radio energy these objects emit, astronomers can learn much about the conditions of stars and galaxies that produce radio energy.

X-rays are another important source of information about the Universe. Fortunately, the Earth's atmosphere is opaque to x-rays, and we must therefore gather them in instruments in spacecraft above the atmosphere. Thus, it is only since the space age that x-ray astronomy has been an important tool for astronomers. X-rays are not gathered in instruments that we would call telescopes, but the devices used to capture x-rays, and gamma rays, are designed for the same purposes as telescopes. That is, they capture as much energy as possible for astronomers to study.

Ultraviolet and infrared wavelengths are parts of the energy spectrum that astronomers have recently started to study in detail. These wavelengths are, respectively, shorter and longer than visible light. Our atmosphere is, again, mostly opaque to ultraviolet radiation, so much of the information in this region of the energy spectrum has been gathered by space satellites. Infrared telescopes have been developed for use on both the surface of the Earth and aboard satellites.

The Magellanic clouds

Our Milky Way Galaxy has two small companion galaxies called the Large Magellanic Cloud and the Small Magellanic Cloud. They are near the South Celestial Pole and therefore visible from any part of the Earth south of 15 degrees north latitude. They are not visible from the United States. They are named for the early sixteenth century Portuguese explorer, Ferdinand Magellan (1480-1521), one of the first explorers to circumnavigate the Earth.

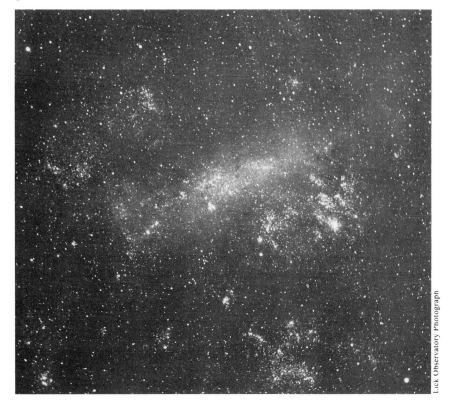

The Large Magellanic Cloud, a small companion galaxy to the Milky Way, is 163 thousand light-years away.

To the unaided eye they appear to be two fuzzy patches of light, easily visible in a clear, dark sky. A telescope reveals a myriad of stars and vast clouds of gas and dust in each galaxy. Neither the Large Magellanic Cloud nor the Small Magellanic Cloud is shaped like the familiar spiral or elliptical galaxies. Thus, they are classified as irregular galaxies, a class that represents only three percent of all known galaxies. Some authorities regard the Large Magellanic Cloud as a barred spiral galaxy.

The Large Magellanic Cloud is thought to be at a distance of about 163 thousand light-years and is moving away from the Milky Way at 34 kilometers per second. The Small Magellanic Cloud is on the order of 200 thousand light-years away and is receding from us at 359 kilometers per second. Due to their proximity to the Milky Way, they are regarded as companion galaxies to our galaxy. Indeed, there is some evidence to show that there is a bridge of hydrogen gas between the Milky Way and the Large Magellanic Cloud.

In 1912, Henrietta Leavitt, working at the Harvard College Observatory, used Cepheid Variable stars in the Magellanic Clouds to establish a period-luminosity relationship (the longer the period, the greater the mean brightness). This relationship in these kinds of stars allowed Harlow Shapley (1885-1972) to determine distances to extra-galactic systems.

In February 1987, a supernova occurred in the Large Magellanic Cloud. This was the closest and brightest supernova visible to the unaided eye to occur in nearly four hundred years.

Mars: Overview

The United States orbited two unmanned spacecraft, Vikings 1 and 2, around Mars in 1976. The principal reason for the mission was to look for life on the planet. Neither spacecraft detected life, but both research packages sent back a large amount of data and information about the fourth planet from the Sun.

Mars is often referred to as the red planet because it appears red to our unaided eye and through a telescope. The color pictures sent back by the Viking spacecraft also show the red surface of Mars, together with a pink sky. There is clear evidence now that the red color is due to a form of iron rust called limonite. Because its red color resembles blood, Mars is named for the Greek god of war.

The reason that scientists are looking at Mars for life beyond the Earth is that the planet more closely resembles the Earth than any other planet in the Solar System. Although its mass is only about one-tenth that of the Earth's mass, and its diameter about one-half of the Earth's diameter (6,800 kilometers compared to 12,756 kilometers for the Earth), Mars rotates on its axis once in 24 hours and 37 minutes, making the Martian day and night only slightly longer than the Earth's day and night.

Furthermore, since Mars's axis is tipped 25 degrees to the plane of its orbit around the Sun, nearly the same as the Earth's 23½ degrees, Mars goes through the same seasons that the Earth does. Mars's seasons, however, are almost twice as long as those on the Earth because Mars requires 687 Earth days to revolve once around the Sun. The Earth takes 365 days.

Because Mars has less mass than the Earth, its gravitation is weaker. If you weigh 150 pounds on the Earth, you would weigh only 57 pounds on Mars.

Mars is an average distance of 227 million kilometers (141 million miles) from the Sun. When the Earth and Mars are on the same side of the Sun, Mars can come as close as 56 million kilometers (35 million miles) to the Earth.

Mars has two small moons, Phobos and Deimos, about 28 and 16 kilometers in diameter, respectively. These satellites were discovered in 1877 by the American astronomer Asaph Hall. Also in 1877, the Italian astronomer Schiaparelli observed streaks on the Martian surface that he named, in Italian, *canali*; this word was unfortunately translated into English as canals. Many people quickly assumed that Mars was inhabited by intelligent creatures. One explanation of the canals was that they brought water from the planet's polar caps to irrigate crops in the temperate zones.

Percival Lowell, an American astronomer, built the Lowell Observatory in Arizona in the early part of this century for the purpose of studying Mars; he was convinced that there was intelligent life on the planet. There were many suggestions made in the late nineteenth century about how to contact Martians. One of the most ambitious ideas involved digging deep trenches in the Sahara Desert, filling them with crude oil, and lighting the oil. The idea was that the Martians would be able to see the signals of these huge fires. Fortunately, this idea was never carried out.

It is important to repeat that neither of the Viking spacecraft detected life on Mars or any evidence to indicate that life may have existed in the past. If life is detected by some future probes, it will quite likely turn out to be subcellular or extremely primitive in nature. We can say with absolute certainty that there is no intelligent life on Mars now and, with almost the same degree of certainty, that there has been no intelligent life on Mars in the past.

Mars: Climate

The United States has sent several spacecraft to Mars, beginning with Mariner IV in 1965, that have either flown by or orbited the red planet and sent back pictures and other data about a variety of surface, atmospheric, and climatological features. In addition, the Viking 1 and 2 missions not only orbited the planet and sent back data and pictures, but instrument packages called "landers" also settled on the surface and sent back many pictures and other information about their landing sites.

These probes, coupled with telescopic observations from the Earth, show us that Mars has an atmosphere made up mostly of carbon dioxide, with traces of oxygen, water vapor, nitrogen, argon, and carbon monoxide. There is so little atmosphere surrounding Mars that the pressure at its surface is equivalent to the Earth's atmospheric pressure at one hundred thousand feet—or more than three times higher than Mount Everest. If human beings are ever to walk on the Martian surface, they will need pressurized space suits.

Future astronauts on Mars will also need heated space suits; the warmest temperature recorded at the Viking 1 landing site was 20 degrees below zero Fahrenheit. The equatorial temperature on the planet in the summer may be more than 90 degrees Fahrenheit, but this would be highly unusual.

The coldest temperatures on Mars are at the planet's north and south poles; they are about 240 degrees below zero Fahrenheit during the winter. At this temperature carbon dioxide in the atmosphere freezes (we call frozen carbon dioxide dry ice); thus, the polar caps in the Martian winter are covered with a thin layer of frozen carbon dioxide. The polar caps are made of water ice as they are on the Earth. They shrink and grow depending on the seasons on Mars. The climate is cold enough that the ice caps never disappear.

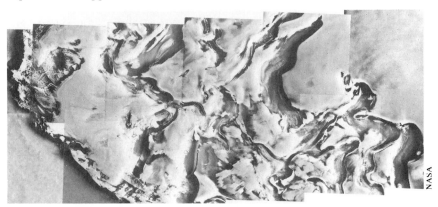

Mars's north polar cap imaged by Viking 2. The bright patches are water ice.

There are also seasonal dust storms on Mars. Strong winds, reaching velocities of three hundred to five hundred kilometers per hour, periodically blow dust from the Martian surface high into the atmosphere. Through Earth-based telescopes these dust storms make the planet appear as if it were undergoing vast surface changes. Indeed, the wind-blown dust piles up in dunes, as revealed by pictures taken by the various Mars probes, and scours other parts of the surface. Wind-blown dust is one of the reasons the Martian surface is eroded.

Mars: Surface features

In the 1970s unmanned space probes to Mars gathered a large amount of information about the red planet. Pictures sent back from Mars by these spacecraft provide evidence that at some remote time in Mars's history there was a great deal of water on the planet. Evidence comes from areas on Mars that appear to have been eroded by flowing water. These areas resemble river channels on the Earth; meanders, islands, cut banks, and other features common to our river systems are clearly seen on the Martian surface.

There is currently no flowing water on Mars. Because the climate of the planet is now quite cold compared, for example, to the Earth's climate, water has been captured in the form of ice at the north and south poles of the planet. Astronomers also think that there must be a large quantity of water frozen just beneath Mars's surface, a circumstance that is similar to the Earth's permafrost in the colder regions of our planet.

Because there is strong evidence that water flowed freely in rivers on the surface of Mars at some time in the past, we conclude that the climate of Mars

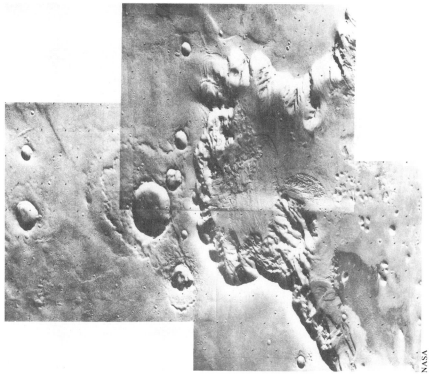

Rock avalanche on Martian canyon floor. Note the sand dunes in the lower right corner of the picture.

was much warmer in the past than it is now. Perhaps Mars had a climate as mild as we are experiencing on the Earth. Now, however, we might say that Mars is in a deep ice age. The reasons for Mars's change of climate and temperature are unknown by astronomers at this time.

Pictures taken by spacecraft in orbit around Mars show craters on the surface of the planet. These craters vary widely in size and age, indicating that Mars has been impacted for a long time by meteors of all sizes. From this evidence we say that Mars generally resembles the Earth's moon rather than the Earth itself.

Another type of surface feature on Mars is enormous canyons. They may have been cut into the surface by flowing water or perhaps were caused by subsurface faulting. One of these canyons is so wide, deep, and long that, if it were on the Earth, it would dwarf the Grand Canyon.

Mars also has several large volcanoes. One of these, Olympus Mons, is 500 kilometers (about 300 miles) in diameter, with a central caldera 70 kilometers in diameter. It is obvious that lava once flowed down the slopes of this

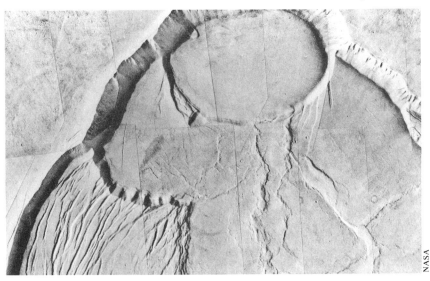

The volcanic crater on top of Olympus Mons is shown from the Viking 1 orbiter. The deepest crater is about 25 kilometers across. At lower left is a lava flow from the caldera down the flank of the volcano.

volcano. All of the volcanoes on Mars now seem to be dormant. There is no apparent reason, however, that they could not become active again in the future. If so, they may supply enough heat to melt at least some of the permafrost, causing water to flow again in the volcanic regions of the planet.

Whenever we send spacecraft to the moon or to the planets to gather information, we always conclude these missions with more questions than we have answered. Mars is no exception. Why, for example, did the climate of Mars change so dramatically? When the climate was less harsh, did the planet support life forms as we know them—or some other type of life? If life existed on the planet, is it now extinct or in some sort of dormant state?

Answers to these and many other questions will undoubtedly be forthcoming in the years ahead as astronomers continue to investigate Mars. In turn, the knowledge we will gain not only will help us to understand the formation and past history of our planet Earth, but also will help us to learn in more precise detail something of the events that created the entire Solar System.

Matter and antimatter

All matter is made up of atoms, which in turn are composed of electrons, protons, and neutrons. Protons and neutrons make up the nucleus of an atom and contain nearly all the mass of the atom; electrons surround the atom's nucleus. In an electrically neutral atom, the number of electrically positive protons equals the number of electrically negative electrons. Neutrons, as indicated by their name, are electrically neutral particles.

The least massive atom is the element hydrogen, composed of one proton in the nucleus surrounded by one electron. If there is a neutron in the nucleus of a hydrogen atom, the atom is said to be an isotope of hydrogen and, in this case, is called deuterium. Hydrogen is by far the most abundant element in the Universe. Helium is the second most abundant element and is also the second lightest element, having two protons in its nucleus and two electrons surrounding it.

The other 90 naturally occurring elements make up a very small fraction of the rest of the mass of the Universe. The heaviest naturally occurring atom is uranium; it has 92 protons and 92 electrons. In its most common form, $^{238}_{92}$U, in addition to the 92 protons, the nucleus contains 146 neutrons.

The nuclei of atoms are divisible not only into protons and neutrons, but also into other kinds of particles that are unstable and decay back into protons and neutrons in a matter of millionths or even billionths of a second. More than three hundred of these subnuclear particles have been identified. Recent theoretical and observational work by physicists suggests that all subnuclear particles may be composed of even more fundamental particles that have been named quarks.

Every nuclear particle has an antiparticle, called antimatter, identical to the particle of regular matter in every way—mass, size, and so forth—except that it has the opposite electrical charge. For example, there is the positron, an electron with a positive electrical charge, and the antiproton, which has a negative electrical charge. An antineutron exists; it decays into an antiproton and a positron.

If a particle of regular matter comes into contact with a particle of antimatter, the result is the conversion of the mass of each particle into energy. This annihilation of mass is the most efficient mass-to-energy conversion known.

There is no way to look at a star or a distant galaxy and say whether it is made of regular matter or of antimatter because antimatter emits radiation in the same way that the regular matter of the Sun does. We know, however, that our moon and the planets Mars and Venus are composed of regular matter because we have landed either spacecraft or instrument packages made of regular matter on each of these objects, and there was no explosion to indicate the conversion of mass to energy. We can probably safely assume that the Sun, the other planets and their satellites, together with the meteoric and cometary material—indeed, all the matter that makes up our Solar System—is regular matter. It has even been suggested that our galaxy is probably nearly all regular matter.

Several astronomers have put forth the notion that the amount of mass in the Universe may be equally divided between matter and antimatter. This idea has a certain aesthetic appeal, but there is no evidence yet to support the speculation.

Mercury

Named after the Roman god who was the fleet-footed messenger for the other gods, the planet Mercury has the distinction of being the second smallest planet in the Solar System (4,878 kilometers or just a bit over 3,000 miles in diameter, only half again as big as our moon). The planet also travels around the Sun more rapidly than any other planet, requiring only 88 Earth days to make one Mercurian year, because it is the closest planet to the Sun at a mean distance of 58 million kilometers. Because Mercury is so close to the Sun, it is only occasionally visible to the unaided eye low in the western sky just after the Sun has set or low in the eastern sky just before the Sun rises.

Its mass is only 1/18 that of the Earth's, and it has an average density of 5.4 grams per cubic centimeter, slightly less than that of the Earth. The planet rotates on its axis in 59 Earth days, or three times for every two revolutions it makes about the Sun. Mercury has no moons.

Mercury is too small to have an appreciable atmosphere, and because it is so close to the Sun without any insulating atmosphere, the surface temperature reaches 700 degrees Kelvin (about 800 degrees Fahrenheit) at noon. This temperature is hot enough to melt some of the softer metals such as tin and lead. During the night on Mercury, the temperature may drop to 90 degrees Kelvin, or 300 degrees below zero Fahrenheit. Therefore, one of the hottest surface temperatures and one of the coldest surface

Southern hemisphere view of Mercury. Note the many impact craters and bright rays caused by secondary impacts.

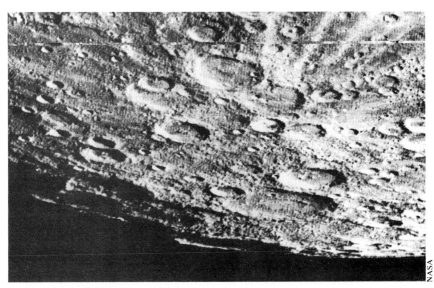

Mercury's south pole imaged by Mariner 10. The largest craters shown in this picture are 200 kilometers in diameter.

temperatures of any of the planets in the Solar System are found on Mercury.

The surface of Mercury, whose details are available to astronomers from the Mariner 10 space probe, resembles the surface of our moon. Thousands of craters—together with cliffs, ridges, scarps, and plains—make up the surface of the planet. It is clear that most of the craters were formed by impacts from meteorites. The existence of plains indicates that there have been large lava flows in the past history of the planet.

Meteors

Most of us have seen streaks of light in the nighttime sky that we have called, erroneously, shooting stars, or falling stars. These flashes of light have nothing to do with the stars. They are small particles of rock and metal that are members of the Solar System; they orbit the Sun just as the planets do. If this material is in space, we call it a meteoroid; if it is in our atmosphere, it is called a meteor; and if it lands on the Earth, it is a meteorite. We see meteors as streaks of light because they are moving fast enough relative to the Earth's motion to burn due to friction with the molecules of gases in our atmosphere. Nearly all meteors burn up between 130 and 80 kilometers above the surface of the Earth.

Most meteors are smaller than pebbles. An occasional one will be larger; when it burns in our atmosphere, it may be large enough to be called a fireball, or bolide. Sometimes bolides can be heard as well as seen. Infrequently, they are seen to shatter into many meteors as they come through the atmosphere.

Meteors that burn in our atmosphere are traveling at speeds that range from 12 to 72 kilometers per second. At a slower speed they would not burn; at a faster speed they would not remain in the Solar System, since 72 kilometers per second is the escape velocity for the Solar System at the Earth's distance from the Sun. The faster a meteor moves through the atmosphere, the brighter it will be. Generally, a meteor that is moving rapidly through the atmosphere will be bluer in color; one that is moving more slowly will be yellower.

More meteors are seen after midnight than before, and they tend to be brighter and bluer. The reason is that before midnight, the Earth's direction of rotation is the same as its direction of revolution about the Sun. The Earth therefore has to overtake any meteoroids in its vicinity, and the speeds of the meteors as they burn in our atmosphere are consequently slower. Many meteors do not travel fast enough to burn, and we never see them.

After midnight, the Earth's direction of rotation carries it into its direction of revolution, and meteors tend to run into our atmosphere. They have a velocity that is the sum of the Earth's rotational speed and of its orbital speed. More meteors are above the minimum 12 kilometers per second speed needed to burn, and we thus see more meteors burning in the atmosphere.

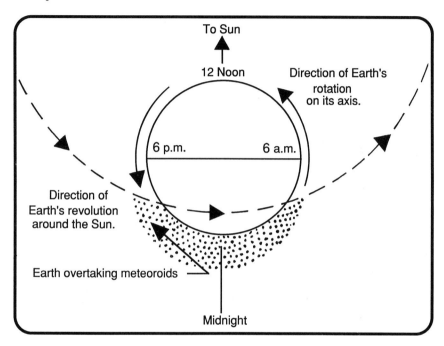

It is estimated that 25 million meteors that are bright enough to be seen with the unaided eye burn in our atmosphere every 24 hours. Of course, we do not see most of them due to daylight, a bright moon, or city lights. It is also estimated that 10 to 100 tons of meteoric material accumulate every day on the Earth. Most of it falls into the oceans, however, since 74 percent of the planet is covered by water.

Meteors: Meteorites

When meteors hit the surface of the Earth, they are called meteorites. There are two kinds of meteorites: stony, which are the most common, and metallic. The metallic meteorites are mostly iron and nickel. The stony meteorites are generally as old as the Solar System itself: their maximum age has been measured at more than 4.5 billion years.

The largest meteorite ever found still intact weighs an estimated 45 tons. It is located in Southwest Africa. The largest meteorite to be moved was found in Greenland by Robert Peary in 1897. It weighs 31 tons and is on display at the American Museum of Natural History in New York.

When large meteorites crash into the surface of the Earth, they often leave craters. One of the largest known meteorite craters in North America is the Barringer Meteor Crater near Winslow, Arizona. It is 1,300 meters in diameter and 180 meters deep. It is estimated that the meteorite that made the crater weighed more than one hundred tons. Apparently it shattered into innumerable fragments on impact because there is no meteorite buried underneath the floor of the crater. The meteorite impacted between 20 thousand and 30 thousand years ago.

There undoubtedly have been untold numbers of large meteorites that have made craters on the Earth during the more than 4.5 billion-year history of our planet. Due to weathering processes, however, craters erode rapidly (in a few tens or hundreds of thousands of years, a very short period in astronomical and geological time scales), and we therefore see only a few of the more recent impact craters.

There is no record of anyone ever having been killed by a meteorite. In 1938, a woman in Illinois recorded a meteorite that crashed through the roof of her garage and imbedded itself in the seat of her car. In Alabama in 1954, a woman was struck, but not seriously injured, by a meteorite that came through the roof of her house and bounced around a room before hitting her.

Meteors sometimes occur in swarms of material that the Earth runs into about the same place in its orbit each year. This material is often associated with a comet that has orbited the Sun and left debris behind that we see as meteors. Meteors in swarms of this kind are called meteor showers. They seem to come from the same place in the sky and are named for the constellation in that part of the sky from which they seem to originate.

Infrequently, a shower can be spectacular. The Leonid shower in 1833 was so intense that observers estimated two hundred thousand meteors an hour could be seen (about 50 per second) for six hours; they were described as appearing to be like a heavy snowfall (see chart p. 91).

Meteor Showers

Shower Name	Approximate Date of Shower Maximum	Hourly Rate	Velocity (kilometers/second)
Quadrantids	Jan. 3	40	41
Lyrids	April 21	15	48
Aquarids	May 4	20	64
Aquarids	July 30	20	40
Perseids	Aug. 11	50	60
Orionids	Oct. 20	25	66
Taurids	Oct. 31	15	28
Leonids	Nov. 16	15	72
Geminids	Dec. 13	50	35
Ursids	Dec. 22	15	34

The Milky Way

On clear, dark summer evenings most of us have seen a faint band of fuzzy light stretching high overhead across most of the sky from north to south. We are looking at the Milky Way Galaxy, a huge collection of stars including the Sun. Since we are a part of the Milky Way, we will never see it from the outside: our view will always be from the Earth or from nearby our planet, and the Galaxy will thus always appear to us as a faint band of light in the nighttime sky. If we use our imaginations and project ourselves outside our

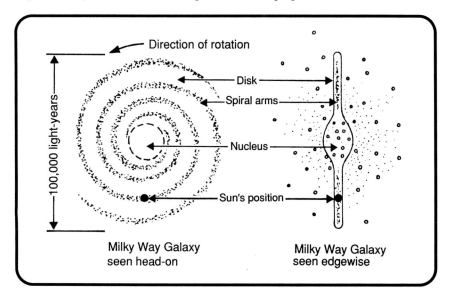

Direction of rotation

Disk

Spiral arms

Nucleus

Sun's position

100,000 light-years

Milky Way Galaxy
seen head-on

Milky Way Galaxy
seen edgewise

galaxy, it would appear to us in the shape of a wagon wheel. The central nucleus of stars is the hub of the wheel, and the arms of the Galaxy are its spokes.

Recent evidence suggests that the Milky Way has a rather active nucleus that provides energy in all wavelengths, from radio to gamma rays. One interpretation is that there is a massive black hole (several million solar masses?) in the center of our galaxy that is the source of energy for all this radiation.

Generally speaking, the nucleus of the Milky Way, composed of billions and billions of stars, contains older stars and little gas and dust. The arms of the Galaxy are younger regions of the system that are made up of stars and large amounts of interstellar gas and dust. It is in these younger regions of the Galaxy where new stars are still being formed. At the age of 4.6 billion years, the Sun is a comparatively young star.

Due to interstellar gas and dust, we are not able to count all the stars in the Milky Way Galaxy. Astronomers make estimates, however, of the number of stars in our galaxy and tell us that there are probably in the neighborhood of four hundred billion stars. The diameter of the Milky Way is more than one hundred thousand light-years; traveling at the speed of light, 186 thousand miles per second (300 thousand kilometers per second), a beam of light would take more than one hundred thousand years to cross the Milky Way.

The Sun and Solar System are located in one of the arms of the Galaxy about 28 thousand light-years from the nucleus. The entire Milky Way is rotating about the nucleus of stars, thus giving it the flattened wheel-like shape. The Sun and the Solar System are being dragged around the nucleus of the Galaxy at a speed of about 225 kilometers per second; at this speed, it takes the Sun approximately 250 million Earth years to make one revolution around the nucleus. All the stars we see in the nighttime sky, whether or not we are looking toward the band of the Milky Way, are in the Galaxy.

Molecules in space

Molecules are groups of atoms of either the same element or of different elements that make up chemical compounds. For example, if two hydrogen atoms are united, we have a molecule of hydrogen gas. If two hydrogen atoms unite with one oxygen atom, the result, H_2O, is the familiar molecule of water.

Using radio telescopes and other detecting devices, astronomers have discovered many kinds of molecules in the spaces between the stars in the Milky Way Galaxy.

Energy from hydrogen gas that exists between the stars was first detected in 1951. Since that time many other molecules have been found in interstellar space that emit enough energy to be detected. Molecules of carbon in combination with both nitrogen and hydrogen are known. If carbon forms a bond with hydrogen, we have a hydrocarbon, a gas similar to those we use for fuels on our planet. Carbon has also been found in combination with oxygen, showing that carbon monoxide exists in space. Water molecules have also been found in interstellar space.

About one hundred different kinds of molecules are known to exist between the stars. In addition to those noted, ammonia, hydrogen sulfide (a gas that smells like rotten eggs), and even more complex molecules have been found. These very complex molecules include formaldehyde (a molecule made of hydrogen, carbon, and oxygen), which is used as a disinfectant and a preservative; hydrogen cyanide (hydrogen, carbon, and nitrogen), which smells like almonds and is poisonous; and methyl alcohol, which is sometimes used as antifreeze in automobiles.

Large molecules that have complex structures of carbon, oxygen, hydrogen, and nitrogen have been found; these molecules provide the basic compounds for amino acids, which in turn are the building blocks of proteins. Proteins are necessary for the formation and maintenance of life forms as we know them.

Because these complex molecules that are the initial compounds necessary for life have been found to be part of the interstellar medium does not mean that there is life in space. However, if we can study the processes by which they are formed in space, perhaps we can learn something about how life originated on the Earth.

The moon: Motions

Originally, the words moon and month were the same words because it takes the moon one month, about 29½ days, to travel once around the Earth. Put another way, the moon moves from one of its phases and back to the same phase (for example, from new moon to new moon or from full moon to full moon) in just a little more than 29 days.

The drawing on p. 94 shows why we see the moon going through its monthly phases. The outer series of moons illustrates the phases of the moon as we see them from the Earth.

Since the moon in first quarter is one-fourth of its way around the Earth, it takes about seven days (one week) to move from new moon to this phase (four divided into 29 is just a little more than seven). Thus, for the moon to move through all four of its phases (new, first quarter, full, third quarter, and back to new again) requires about four weeks, or one month.

The moon rises an average of 50 minutes later each day because it revolves around the Earth from west to east, the same direction that the Earth turns on its axis. It takes any point on the Earth this extra amount of time, 50 minutes, to catch up with the moon from moonrise to moonrise. Of course, the moon also sets an average of 50 minutes later each day, too.

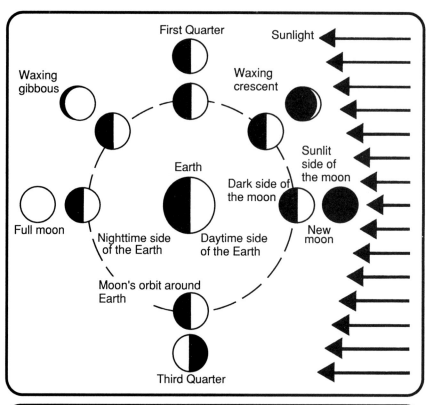

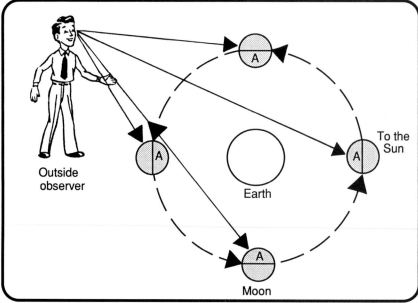

Because the moon always keeps the same side facing the Earth—except in pictures sent back from the far side of the moon, no one on Earth has ever seen the moon's back side—we say that the moon rotates once on its axis for every revolution it makes around the Earth. On Earth, we always see side A of the moon. An outside observer, however, would see all sides of the moon in one month, and thus would say that the moon turns, or rotates, once on its axis for every trip it makes around the Earth (see drawing p. 94).

Strictly speaking, the moon does not revolve around the Earth, but rather the Earth-moon system swings around a common center of mass. Since the Earth is 81 times more massive than the moon, however, the common center of mass of the two objects is close to the center of the Earth, so close that it is about 17 hundred kilometers beneath the surface of the Earth.

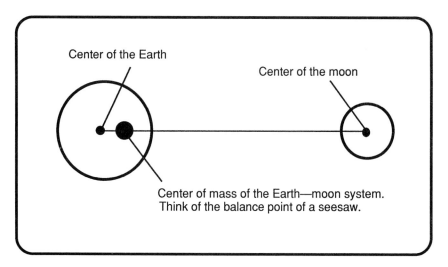

Center of the Earth

Center of the moon

Center of mass of the Earth—moon system.
Think of the balance point of a seesaw.

When the moon is new or full, it is closely lined up in the same straight line with the Sun and Earth. This arrangement is called syzygy, a word you can use to get rid of the letter "z" when playing Scrabble. Sometimes this word appears in crossword puzzles, too.

The moon: Appearance

We see the moon, the Earth's only natural satellite, because it reflects light from the Sun to the Earth. The moon is brightest, of course, when it is full, so bright that we can read a newspaper by its light. Yet, even when it is full and very bright, it sends us only about 1/400,000 the amount of light that the Sun does on a clear day.

Astronomers talk about albedo, the capacity of an object to reflect light. The moon is really a dull object, having an albedo of 0.07; that is, it reflects only seven percent of the sunlight it receives and absorbs the other 93 percent. Because the moon has a mass smaller than the Earth's, its gravitational field is weak, and it is unable to hold an atmosphere. Even the heaviest gases escape from the moon. Consequently, there is nothing but the dull gray surface of the moon to reflect sunlight; the moon appears to be quite lackluster in comparison to the Earth.

The Earth's albedo is 0.35; the Earth reflects 35 percent of the light it receives from the Sun back into space. The reason that the Earth has a much higher albedo is because we have an atmosphere that contains a great deal of water. Sunlight is reflected from clouds made of water vapor, as well as from the oceans, and thus makes the Earth much brighter than the moon.

When we see the small, thin crescent of the new moon setting just after the Sun in the western sky, we can also just barely see the dark side of the moon. Sunlight is reflected from the Earth to the dark side of the moon and back to the Earth again, giving us a shadowy glimpse of the nighttime side of the moon.

The moon is 3,476 kilometers in diameter, about one-fourth the diameter of the Earth. Its average distance from the Earth is 384,400 kilometers (about 240,000 miles); it can be a maximum of 406,600 kilometers and a minimum of 356,300 kilometers from the Earth. The difference is due to the

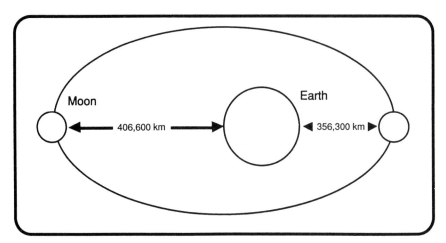

nature of the moon's orbit around the Earth. It is not a perfect circle but rather a slight oval shape called an ellipse. The moon thus appears larger when it is closer than when it is farther from the Earth during the same phase.

Another commonly observed phenomenon that has to do with the size of the moon is the apparent size of the full moon when it rises over the eastern horizon. It seems to be much greater in size when it is close to the horizon than a few hours later when the full moon is high in the sky. Because the full moon is farther away from us when it is rising than when it is high in the sky, it should appear smaller to us by an amount that is detectable by the human eye. Why the reverse situation is true no one has yet been able to explain except to suggest that the apparent larger size of the rising full moon is a psychological rather than a physical phenomenon.

The moon: Surface features

When Galileo looked at the moon through the first astronomical telescope he saw mountains, craters, and large smooth areas which he mistakenly named *maria*, the Latin word for seas. Today we know that there is no water on the moon; what Galileo saw we recognize as large, relatively flat and smooth areas on the moon that are better described as plains.

The plains are clearly the result of molten rock, although the origin of this molten material is not yet understood. If the moon at one time had a hot core, molten rock may have oozed onto the surface. Alternatively, heat from radioactive decay in the outer layers of the moon may have been sufficient to melt the surface material.

It is interesting to note that only the side of the moon that faces the Earth has large *maria*; the far side of the moon is free of this type of lunar surface feature.

Craters are also common features on the moon. The largest craters are about 240 kilometers in diameter. The *maria* seem to have fewer craters than other parts of the moon, indicating that they may be younger. The largest craters are called walled plains. The inside walls of the largest craters can rise to three thousand meters above the craters' floors. Many craters have peaks in their centers. Craters are almost certainly caused by impacts of meteors on the surface of the moon during its several-billion-year history.

There are several mountain ranges on the moon, too, many of which have been named after mountain formations on the Earth; thus, we have the lunar Alps and the lunar Apennines. The highest mountains on the moon are perhaps eight thousand meters—just a bit higher than Mount McKinley in Alaska—with the average size probably about half as tall.

Lunar valleys are also prominent; they seem to be gorges in the surface of the moon, perhaps the result of faulting in the moon's crust. Rays are bright streaks that radiate from some craters. These rays may be the result of material splashed from the craters and can thus be thought of as strings of secondary impact craters.

The surface of the moon is also covered with dust, the result of the accumulation of fine meteoric material during billions of years. Dust may

Lick Observatory Photograph

The full moon photographed with the 36-inch refractor at Lick Observatory. The large crater in the south (bottom) is Tycho, about 96 kilometers in diameter.

also be created when meteors impact the surface, pulverizing rocks as they hit the moon's surface.

Because there is no atmosphere on the moon, there are no erosional processes similar to those on the Earth. The lack of water on the moon means that the oldest surface features, such as the mountains, that we see through telescopes have been there for billions of years. All the surface features on the moon are thus much more permanent than they are on the Earth.

Other results of the lack of a lunar atmosphere are temperature extremes and rapid temperature changes. Rocks in direct sunlight on the moon's surface have a temperature above the boiling point of water (373 degrees Kelvin). These same rocks when plunged into shadow will achieve temperatures in a matter of a few hours that may reach as low as 100 degrees Kelvin. Temperatures can drop 150 degrees in one hour.

Surface rocks on the moon are made up of basalts similar to those on the Earth. The most common elements found on the lunar surface in various kinds of mineral compounds are oxygen in combination with silicon, and aluminum, calcium, iron, magnesium, and titanium.

The origin of the moon is not yet clear. The most popular current theory is that early in the history of the Solar System the Earth was hit by a Mars-size object that knocked a significant amount of material from our primitive planet. This material subsequently went into orbit around the Earth and, through the processes of chemical differentiation and bombardment, evolved into the moon we know today.

Moons of the planets

There are 63 natural moons, or satellites, that we know of going around seven of the nine planets in the Solar System. We know, of course, that the Earth has one moon. Our moon, even though it is not the largest satellite in the Solar System, is the second largest moon in comparison to its parent planet. This circumstance gives rise to the notion that the Earth and moon may be referred to as a double planet system.

Six of the satellites in the Solar System are larger than the Earth's moon. Nearly all of the moons in the Solar System go around their parent planets in a west to east direction, as our moon does, but several go in the opposite direction. Most satellites in the Solar System also follow the equatorial planes of their parent planets; the moon is an exception, following more closely the plane of the Earth's orbit around the Sun. The mass of all the moons put together is only 4/100,000 of the total mass of the Solar System. (About 99.86 percent of the mass of the Solar System is contained in the Sun.)

Jupiter has 16 moons. Four of the five moons closest to the planet were discovered by Galileo with the first astronomical telescope in the beginning of the seventeenth century. They are bright enough and large enough to be seen with the unaided eye were it not for the brightness of Jupiter itself.

Jupiter's two largest moons are about five thousand kilometers in diameter, about half again as large as our moon. Two others are about the size of our moon. Jupiter's smallest moon is estimated to be only eight kilometers in diameter. The most distant satellite is 24 million kilometers from Jupiter and takes about two Earth years to revolve once around the parent planet.

None of the moons of Jupiter has a detectable atmosphere. The four outer moons revolve from east to west around the planet, giving credence to the notion that they may have been captured by Jupiter rather than formed from the debris left over from the formation of Jupiter itself.

If we think of Jupiter as an object that is almost a star, we can then regard its 16 satellites as resembling planets. Thus, Jupiter and its satellites may be thought of as a smaller version of the Solar System.

Saturn has 20 known moons, in addition to its rings, the largest of which is named Titan. This moon, larger than our own, is also one of only two moons in the Solar System known to have a permanent atmosphere. (Neptune's moon, Triton, is the other moon that is known to have a permanent atmosphere.) Nitrogen is the most prominent gas in the atmosphere of Titan.

Mars has two tiny moons, Phobos (the Greek word for fear) and Deimos (panic), 28 kilometers and 16 kilometers in their longest dimensions respectively. Phobos revolves from west to east around Mars in seven hours and 39 minutes. However, since Mars rotates on its axis in just a little more than 24 hours, Phobos appears to rise in the west and set in the east, as seen from the surface of the planet, just the opposite of our moon.

Space probe pictures of Phobos and Deimos show them to be cratered, pitted, and desolate moons. It is interesting to note that Johannes Kepler predicted, on nonscientific grounds and long before Earth-based telescopes powerful enough to spot Phobos and Deimos were invented, that Mars would have two moons. Jonathan Swift, in *Gulliver's Travels*, repeated Kepler's prediction.

Uranus has 15 moons, and Neptune has 8 moons. Neither Mercury nor Venus has any known satellites. Pluto has a moon, Charon, that is larger in comparison to its parent planet than any other moon in the Solar System.

Neptune

The eighth planet in the Solar System is Neptune, named for the Greek god of the sea. It orbits the Sun at an average distance of 4.5 billion kilometers and is so far from the Sun that it takes 165 Earth years to make one complete revolution. Put another way, it will make its first complete orbit around the Sun, since its discovery in 1846, in the year 2011.

The Voyager 2 spacecraft encountered Neptune in August 1989. A great deal was learned about Neptune and its largest moon, Triton. We now know that Neptune rotates on its axis in a little more than 16 hours and that it is 49,528 kilometers in diameter. Its mass is about 17 times the mass of the Earth, and its density is thus 1.64 times the density of water. Voyager 2 also discovered an Earth-sized Great Dark Spot in the upper atmosphere of Neptune, not unlike the Great Red Spot in Jupiter's clouds.

The planet's atmosphere is dominated by hydrogen, but there is also some methane and smaller amounts of helium and ammonia. The minimum temperature at the tops of the clouds is 50 degrees Kelvin (370 degrees below zero Fahrenheit). Neptune must thus have an internal heat source because it emits 2.7 times as much energy, mostly in the infrared, as it receives from the Sun.

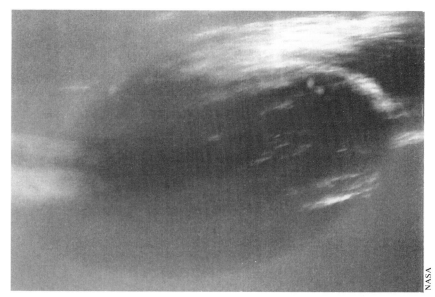

NASA

Great Dark Spot in the upper atmosphere of Neptune. This Earth-sized disturbance resembles the Great Red Spot in Jupiter's atmosphere.

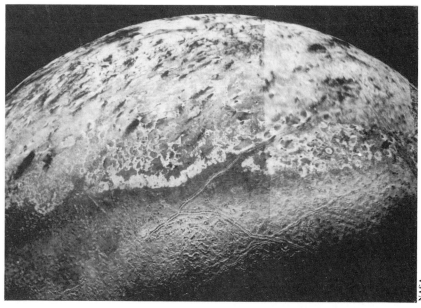

NASA

Image of Triton taken by Voyager 2. Triton is the second coldest major object in the Solar System. Its geologically young surface is constantly being reshaped by apparent outgassing, winds, deposits of frozen material from the atmosphere, and perhaps ice flows.

Neptune has eight moons, one of which, Triton, is among the most interesting objects in the Solar System. It has the lowest surface temperature of any object in the Solar System, 38 degrees Kelvin (391 degrees below zero Fahrenheit). Its atmosphere is mostly nitrogen, together with a small amount of methane. There are few craters on Triton, indicating a surface that, in geological time scales, is constantly being reworked. It is so cold on Triton that most surface features are covered with a thin layer of frozen nitrogen and ammonia.

Triton orbits in a retrograde direction, or clockwise sense, around Neptune; that is, it orbits from east to west, the opposite direction from its other seven moons. This circumstance leads astronomers to believe that Triton was formed elsewhere in the outer Solar System and gravitationally captured by Neptune. It is thought that Triton and the planet Pluto resemble each other.

The Voyager 2 spacecraft also discovered four rings around Neptune. We now know that all the giant planets in the Solar System—Jupiter, Saturn, Uranus, and Neptune—have rings around them.

The discovery of Neptune in 1846 was a triumph of precise, predictive, mathematical astronomy. In the early part of the nineteenth century it was known that the seventh planet from the Sun, Uranus, was moving in an orbit that showed slight but significant deviations from its predicted path. Two mathematicians, John Couch Adams in England and Urbain Leverrier in France, independently suggested that these deviations were the result of the gravitational pull of an unknown planet; both predicted where this planet should be. It was found by the astronomer Johann Galle, at the Berlin Observatory, less than one degree from where Leverrier predicted it would be. (Adams predicted its position within two degrees.)

We sometimes have to think of Neptune as the farthest planet from the Sun. Due to the slightly elliptical (oval) orbits of both Neptune and Pluto (the planet we usually regard as being the last planet in the Solar System), Neptune is occasionally farther from the Sun than is Pluto. This is the situation from 1977 to 1999.

Isaac Newton

Isaac Newton was born in Woolsthorpe, England, on Christmas Day, 1642, the same year in which Galileo died. Newton's father was a small landowner and farmer, who died before Isaac was born. Although he did not display any particular mental capability as a youth, young Isaac had a good facility with his hands, an ability that was important to him in his scientific career.

He entered Trinity College, Cambridge University, in 1661 and soon displayed the mathematical genius that was to make him the most famous scientist of his time. In 1665, and again in 1666-1667, a plague ravaged

England; the university was closed, and Newton returned home. It was during this period, as a young man, that he accomplished some of his most important work. Newton formulated the laws of motion and the universal law of gravitation in which he demonstrated that every object attracts every other object in the Universe.

It was also during this time that he conducted his famous experiments on light with the prism in which he showed that light behaves as a wave as it is broken into its spectrum of colors. He also invented a new branch of mathematics in order to explain his interpretation of the motion of objects in space. He called his new mathematics "fluxions"; today we know it by the more familiar name of the calculus.

In 1668, Newton, at age 26, became Lucasian Professor of Mathematics at Trinity College. He was elected, in 1672, to the Royal Society of London, the most prestigious scientific organization in England, largely on his invention of the reflecting telescope. All large, modern, optical telescopes are variations of Newton's invention. During this period Newton also carried out many chemical experiments.

At the urging of his friends, particularly Edmund Halley (after whom the famous comet is named), Newton published, in 1687, many of his mathematical and astronomical ideas. The title of this monumental book is *Mathematical Principles of Natural Philosophy* (in Latin, *Philosophiae Naturalis Principia Mathematica*—in Newton's day nearly all learned works were published in Latin, the universal language understood by educated people). This book developed precise statements about the orbits of the planets, moon, comets, and other celestial objects, and the motion of objects through the atmosphere and water (Newton thus laid the mathematical foundation for modern engineering including aerodynamics and hydrodynamics); explained the causes of ocean tides and precession of the equinoxes; described the precise shapes of the moon and Earth; and calculated the motion of objects near the Earth (thus giving us the basis for modern rocketry and space travel).

For the first time in human thought, adequate mathematical explanations and proofs were provided for most phenomena that could be observed in the physical Universe in the seventeenth century. Predictability of physical events gained enormous accuracy. It can be said without exaggeration that the *Principia*, as it is known, has had more influence on modern science and thought than any book published since Newton.

In 1689, Newton was elected to Parliament. In the early years of the 1690s he suffered a nervous breakdown but recovered in time to be appointed to the post of Warden of the Mint in 1696; in 1700, he was appointed Master of the Mint. Newton was elected president of the Royal Society in 1703, a position he held until his death in 1727. In 1705, he was knighted by Queen Anne.

For the last 30 years of his life he did no new important scientific work, although he published results of much of his earlier work. During this latter period he was deservedly regarded as the greatest scientist of his era and perhaps of all time.

The North Star

Perhaps the most famous star in the nighttime sky is Polaris, the North Star. Never moving in our northern sky and having no other bright stars nearby, it is a constant guide to the direction north and to our latitude on Earth (distance north of the equator).

The Earth's north axis points almost directly at the North Star. Thus, from our position on the rotating Earth, Polaris seems to stay in the same part of the sky; all the other stars appear to circle around Polaris.

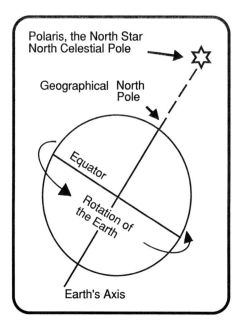

If we were to travel to the north geographical pole, the North Star would be directly overhead and, as the Earth rotates on its axis, all the stars would seem to circle around the North Star, neither rising nor setting. If we are on the equator, the North Star will be right on our northern horizon, and all the stars in the nighttime sky rise in the east and set in the west.

We always know our latitude by the altitude of Polaris above our northern horizon. Obviously, if we are at the equator our latitude is zero degrees north and, as we have seen, Polaris is just on our northern horizon with an altitude of zero degrees.

If we are at the North Pole, Polaris is directly overhead, 90 degrees from any horizon. Thus, our latitude at the North Pole must be 90 degrees north of the equator. The central part of the United States is about 40 degrees north of the equator (40 degrees north latitude); the North Star is thus 40 degrees above the northern horizon.

Contrary to popular belief, Polaris is not the brightest star in the sky; throughout the sky there are 50 stars brighter than Polaris. Furthermore, Polaris changes brightness in a period of just under four days but so slightly that the change is barely detectable by the unaided human eye. It is also a binary star; that is, Polaris is a double star system. The system is about 820 light-years from the Solar System.

Novae

Should a star suddenly become brighter in the sky, it is called a nova (plural, novae). The energy released in the form of visible light and other radiation is not nearly as great as the energy from a supernova explosion. Novae are much more common events than supernovae and are not caused by entire stars exploding, as supernovae are, but by an entirely different mechanism.

Imagine a binary star system; that is, two stars revolving around a common center of mass. One of the stars might be a white dwarf, an old, compact object that has gone through nearly all of its lifetime. The other star is probably younger with an extended atmosphere, such as a giant or supergiant star. The companion star could also be a hot star that has a substantial stellar wind of material blowing away from it.

In either event, material from the companion star accumulates on the surface of the white dwarf. The material may hit the surface of the white dwarf at a substantial speed. A high temperature results, which leads to an explosion as the mostly hydrogen gas from the companion star fuses into heavier elements at the surface of the white dwarf. Alternatively, if the gas from the companion accumulates more slowly, over hundreds of thousands of years, the temperature will accordingly rise more slowly until it is sufficient to ignite nuclear fusion on the surface of the white dwarf.

A white dwarf has a hot surface, perhaps one hundred thousand degrees Kelvin, but is not very bright because it is so small, perhaps no larger than the Earth. When fusion takes place on its surface, it becomes much brighter than almost all the other stars in the sky.

Material from the companion star either reaches the surface of the white dwarf at regular intervals or builds up at regular intervals. In these cases the novae become brighter at regular intervals; these novae are then called cataclysmic or recurrent novae.

It is also possible that so much matter from a companion star can fall onto a white dwarf that it will be more than 1.4 times the mass of the Sun. In this case, there will not be a nova explosion but a supernova event. The result will be either a neutron star or a black hole.

Observatories

From 1948 to 1976, the largest telescope in the world was located in the Hale Observatory on Palomar Mountain in Southern California. Its primary lens, a mirror, is 200 inches in diameter, nearly 17 feet (5.08 meters). Since 1976, the Russians have operated a telescope that has a mirror 6 meters in diameter (about 236 inches) in the Caucasus Mountains.

The Earth's atmosphere causes stars to twinkle; while this effect may be attractive in nursery rhymes ("Twinkle, twinkle little star, how I wonder what you are..."), it is disastrous for the astronomer who needs to work with the clearest, sharpest, most steady beam of light possible. Thus, large research observatories are nearly always built on top of high mountains where the "seeing" is good. The mountains are above a substantial part of the Earth's atmosphere; thus, astronomers are able to avoid some of the turbulence in our atmosphere that makes the stars twinkle.

Furthermore, large cities, which produce a great deal of light that competes with the light from faint astronomical objects, are not usually found in high mountains. Astronomers want to be as far away as possible from the pollution of city lights: auto headlights, street lights, factory, advertising, and commercial lights. Indeed, most of us in the United States live in urbanized areas flooded with light and thus have lost our perception of the nighttime sky. Life might be more rewarding for us if we had less artificial light at night and could once again become familiar with stars, planets, and the moon.

The demand for observing time by research astronomers on any large telescope far exceeds the amount of time available. Therefore, observatories are built in locations that have not only good seeing but also a large number of nights during the year that are cloudless so that as many astronomers as possible can be accommodated on the telescopes. The number of sites that combine all of these features—good seeing, a large number of clear nights, and minimum interference from artificial lights—are becoming scarce on our overcrowded planet.

Large observatory facilities in the United States include the Mount Wilson and the Lick observatories in California, the Kitt Peak and Lowell observatories in Arizona, the McDonald Observatory in Texas, and the Yerkes Observatory in Wisconsin, which houses the world's largest refracting telescope with a lens diameter of 40 inches. The highest large observatory is about 13 thousand feet, located in Hawaii. The oldest college observatory in this country was established in 1836 at Williams College in Massachusetts.

Within the past two decades large observatories have opened in the Southern Hemisphere, notably in Chile and in Australia, in order that astronomers can study the stars, nebulae, and galaxies of the southern skies.

Palomar Observatory Photograph

The 200-inch telescope at Palomar pointing toward the zenith.

Orbits

Everything in the Universe is in motion. Moons orbit planets, planets orbit the Sun and presumably other stars, the Sun and all the other stars in the Milky Way orbit the center of our galaxy, galaxies in clusters move in complex orbits about the center of mass of their clusters, and clusters of galaxies move away from each other in the general expansion of the Universe. All of these motions involve the concept of an orbit.

The ancient Greeks believed that the motions of the planets were perfect circles. Today we know that there is no such orbit in the Universe: perfect circular motion does not exist. There are really just two kinds of orbits: closed orbits and open orbits. Closed orbits are displayed by moons going around the planets, planets moving around the Sun, stars moving around the centers of their galaxies, and galaxies moving around the centers of mass of their clusters. These orbits are said to be closed because the objects that move in closed orbits are gravitationally bound to a larger mass. Thus, moons are bound to planets, planets to the Sun and other stars, stars to galaxies, and galaxies to clusters of galaxies. These closed orbits are all ellipses, or oval shaped figures.

The amount of "ovalness," or flattening of the ellipse, is called the eccentricity of the ellipse and is the ratio, in the case of the Sun and Earth, of

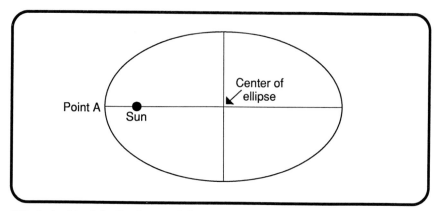

Elliptical orbit of the Earth around the Sun, highly exaggerated.

the length of the line from the Sun to the center of the ellipse divided by the length of the line from point A to the center of the ellipse. Obviously, if this number is zero (i.e., the Sun is at the center of the ellipse), we have a special case of the ellipse called a circle. Since the Sun can be at only one place relative to the Earth (or another planet) in a circle, whereas in an ellipse the Sun can be at an infinite number of points from the center of the ellipse to point A, the chances of any orbit being an ellipse compared to a circle are infinity to one. This is the reason there are no circles in the Universe. All closed orbits are thus ellipses. The Earth has a nearly circular orbit around the Sun; Pluto and Mercury have the most eccentric orbits (orbits that depart from a circle the most).

Orbits are said to be open when one object encounters another object just once. For example, the Voyager 1 and 2 spacecraft, launched from the Earth,

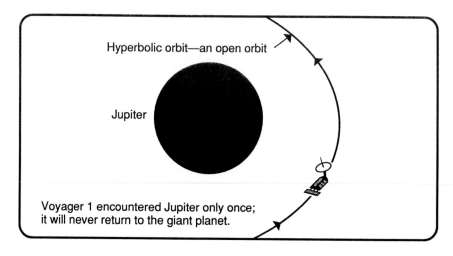

each encountered Jupiter and Saturn once and then sped off to the next planet and eventually out of the Solar System.

A closed orbit that is infinitely long, and would thus appear to be open, is said to be parabolic and does not exist in the Universe for the same reason that a circle does not exist. There can be an infinite number of hyperbolic orbits of Voyager 1 around Jupiter, for example, but only one parabolic orbit. Thus, the chances of a hyperbolic orbit, compared to a parabolic orbit, are infinity to one. Hence, all open orbits are hyperbolas.

Parallax and parsecs

Hold your index finger at arm's length and look at it against some busy background, books on a bookshelf for example, first with just one eye open and then the other eye open. You will see your finger jump back and forth against the background of books. Put another way, because there is a distance between your two eyes, you see your finger against two slightly different parts of the background.

This effect is called parallax, and it is this concept we use to establish the first step in determining the size of the Universe. In June, we will see the

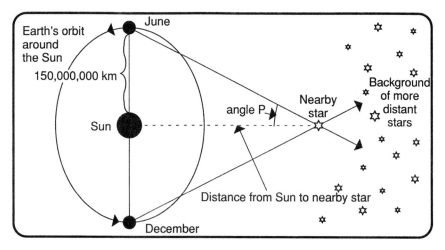

nearby star against a background of more distant stars. In December we will see that the same nearby star will have moved slightly against the background of more distant stars. If we photograph the nearby star at these two times and put the two photographs together, we will have a combined picture (see drawing p. 110).

The image of the nearby star would appear in two places against the background of more distant stars in the same way that your index finger appeared in two places against the background of books when you looked at

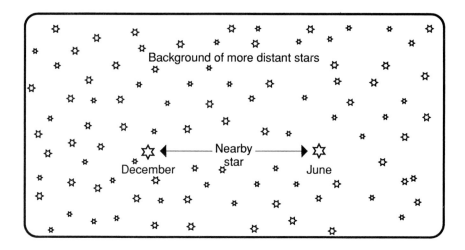

it first with one eye and then with the other eye. The distance between your eyes is analogous to the distance across the Earth's orbit: your index finger is the nearby star, and the books are the more distant background stars. The distance between the two images of the same star, as pictured above, is measured as an angle; one-half of this angle is called the parallax angle (angle p on the first diagram). Knowing angle p and the distance between the Earth and the Sun (about 150 million kilometers), we can easily calculate the distance between the Sun and any nearby star whose parallax angle we can measure.

The star with the greatest parallax angle is Proxima Centauri in the Alpha Centauri system; its parallax angle is 0.76 arc second (an arc second is 1/3,600 of a degree), or about the same angle made by the two ends of a foot ruler viewed from a distance of 51 miles. This star is at a distance of about 42 million, million kilometers (26 million, million miles), which is about 4.32 light-years. Since it has the largest parallax angle of any star, it is obviously the closest star. Reliable distances to the closest one thousand stars have been measured in this way. Obviously, the farther away a star is, the smaller (and hence harder to measure) its parallax angle.

Based on the idea of parallax, astronomers have developed a unit of distance called the parsec, which is the distance a star would be if it had a parallax angle of one arc second. There is no star this close, but if there were it would have a distance of about 30 million, million kilometers (19 million, million miles), or 3.26 light-years.

It is important to know the accurate distances to the nearest stars because we can then compare their characteristics (brightness, for example) with more distant stars, both in our own galaxy and in more remote galaxies, to give us a way of calculating distances throughout the Universe.

Photometry

Light can be thought of as behaving both as waves and as particles. When we analyze the wave nature of light—divide it into its various wavelengths—we can determine the physical characteristics that produced those wavelengths (temperature, element composition, element abundance, motion), and we are doing spectroscopy. When we analyze the particle nature of light, we are doing photometry. Particles of light are referred to as photons; hence, photography means recording photons, and photometry means measuring photons.

What we do when we measure photons is measure their energy content, which will give us accurate information about the temperature of a star. If we heat a piece of iron to a low temperature, it will be a dull red. At succeeding higher temperatures it will be orange, yellow, and finally blue-white. The same temperature, or color, sequence is true of stars, and we use this fact to analyze their temperatures.

Suppose we attach a photometer, a device for measuring the brightness of a star, to a telescope. Then we measure the brightness of a star in three different parts of the spectrum: near ultraviolet, blue, and visual (yellow). We abbreviate this photometric system as UBV. If the star is very hot, it will put out more energy (i.e., be brighter) in the near ultraviolet part of the spectrum compared to the blue part of the spectrum and certainly will put out more energy (be brighter) compared to the yellow part of the spectrum.

If a star is not quite so hot, it will be brighter in the blue part of the spectrum than either the near ultraviolet or the yellow. If a star is cooler, such as the Sun, most of its energy and hence brightness will occur in the yellow part of the spectrum.

With suitable filters and a detecting device such as a photocell, the brightness of stars—their energy output—in various parts of the spectrum can be measured accurately, and hence their temperatures can be determined. Photometry is thus an important analytical tool complementary to spectroscopy. One of the most common comparisons of photon energy is the one we have discussed, the UBV photometric system.

Planetary nebulae

For most of its lifetime a star is in equilibrium; that is, it maintains a balance between the flow of energy from hydrogen fusion processes in its core and gravity that tries to collapse the star in on itself. When a star about the mass of the Sun has fused all of the hydrogen in its core into helium, the star is near the end of its life. Energy can no longer flow from its core, and the tremendous crush of gravity pulling all the outer layers of gas onto the core of the star causes the star to collapse.

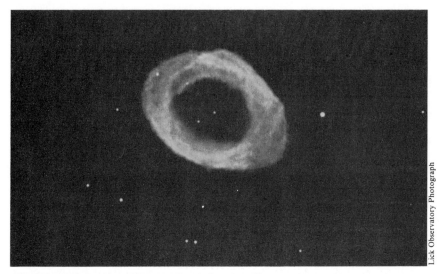

The Ring Nebula, an exploding star, in the constellation of Lyra. The white dot in the center of the nebula is a white dwarf, the highly condensed and dying core of the star.

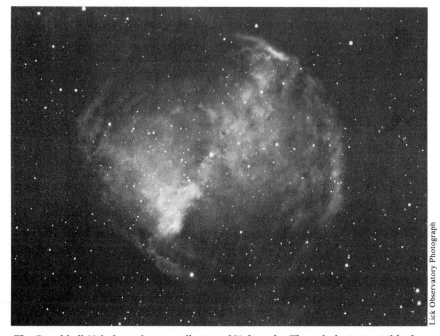

The Dumbbell Nebula in the constellation of Vulpecula. The nebulosity is visible due to ultraviolet light from the hot white dwarf (central dot) energizing the gas in the expanding shell.

As the star collapses, temperature and pressure rise in the core region. The outer regions of the star are pushed away from the core by a new flow of energy. The energy comes from the helium core fusing into carbon and from the hydrogen shell around the helium-fusing core also fusing into more helium. The result of these processes is the release of energy in great enough amounts to push the outer layers of gas completely away from the star, thus creating a stellar wind. A substantial amount of the star's mass can be lost via the stellar wind. The star is now said to be passing through its red giant phase into its planetary nebula phase.

When we see these expanding objects through a telescope, they appear as a shell or ring of gas surrounding a hot dense core called a white dwarf. The expanding ring of gas is visible because ultraviolet light from the white dwarf (whose surface temperature can be one hundred thousand degrees Kelvin or more) energizes its atoms, primarily oxygen, hydrogen, and nitrogen, in the ring and causes them to fluoresce, or to give off their own light.

These objects, a few of which were first described in the latter part of the eighteenth century, resembled faint planets when viewed through early telescopes—hence the name planetary nebula (plural, nebulae), although they have nothing to do with planets. The word nebula is Latin for cloud. We must remember that, to early astronomers in the eighteenth and nineteenth centuries, all nonstellar objects—what today we know as planetary nebulae, galaxies of all types, clouds of gas and dust, and globular clusters—were called nebulae.

Two well-known planetary nebulae, the Ring Nebula in the constellation of Lyra and the Dumbbell Nebula in Vulpecula, are visible through a small telescope in the summer sky (see photos p. 112).

The Ring Nebula is five thousand light-years from the Solar System. Its shell, or ring, is expanding at several tens of kilometers per second, but because the star is so far away, we cannot see the ring expand. The Ring Nebula will thus appear virtually the same for hundreds or perhaps thousands of years. It is thought to have begun expanding 10 thousand years ago.

We observe a ring in many planetary nebulae, even though the material from the outer layers of the white dwarf star is expanding away from the core in all directions. The reason we see an apparent ring is that we are looking through more material at the edge of the expanding mass of gas than when we look at the center (see drawing p. 114). The longer line of vision thus allows us to see more matter, and we have the illusion that we are looking at a ring.

The distance across the ring in the Ring Nebula is about 1.7 light-years. Thus, we are seeing what the length of 1.7 light-years is at a distance of 5,000 light-years.

There are estimated to be 20 thousand to 50 thousand planetary nebulae in our galaxy. It is thought that our star, the Sun, will end its life—and, of course, destroy all the planets in the Solar System—as a planetary nebula and finally a white dwarf in another five billion years.

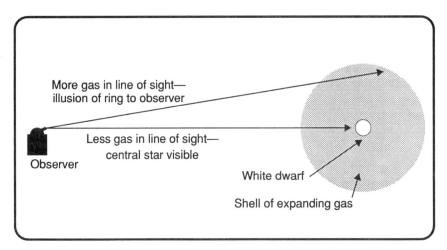

More gas in line of sight—
illusion of ring to observer

Less gas in line of sight—
central star visible

Observer

White dwarf

Shell of expanding gas

Pluto

Discovered in 1930 at the Lowell Observatory near Flagstaff, Arizona, Pluto is the most distant planet that we know about in the Solar System. The search for a planet beyond Neptune was initiated by Percival Lowell, among other astronomers, at the beginning of the century in order to account for small irregularities in the orbit of Uranus, the seventh planet from the Sun. These irregularities turn out not to have been real, but the mistake led Clyde Tombaugh, an American astronomer working with photographic plates taken at the Lowell Observatory, to find Pluto very nearly where Lowell predicted it would be—based on false data. Thus, the discovery of Pluto was accidental.

Pluto is named for the god of the underworld, appropriately enough since it is on the outer fringes of the Solar System where light and warmth from the Sun are very weak. Indeed, as seen from Pluto, the Sun would be merely a bright star in a dark sky. The first two letters of the planet's name, PL, are also the initials of the astronomer, Percival Lowell, who persisted in the search for the planet. Its name thus honors him.

Pluto's orbit is inclined to the ecliptic (the plane of the Earth's orbit) more than the orbit of any other planet. Furthermore, its orbit departs more from a circle—that is, it is more eccentric—than the orbits of any of the other planets. At its farthest, Pluto is more than 7.0 billion kilometers (about 4.3 billion miles) from the Sun; and at its closest, it is less than 4.5 billion kilometers (about 2.7 billion miles) from the Sun. Thus, part of its orbit is closer to the Sun than Neptune's orbit, so we say that most of the time Pluto is the ninth planet from the Sun, but occasionally it is the eighth planet in the Solar System, a situation that obtains from 1977 to 1999.

Because it is the farthest planet from the Sun, Pluto requires more than 249 Earth years to make one revolution around the Sun; the Plutonian year is longer than that of any other planet in the Solar System.

Pluto and its moon, Charon, imaged by the Hubble Space Telescope. This is the clearest picture ever taken of the Solar System's most distant planet.

Pluto, together with its moon, is thought to have a mass of about 1/400 the Earth's mass, a diameter of 2,245 kilometers, and consequently a density about twice that of water. Pluto rotates on its axis every 6.4 Earth days. It probably does not have much of an atmosphere; in any case, most of the gases would be frozen since the temperature is measured at only 40 degrees above absolute zero Kelvin, which is 387 degrees below zero Fahrenheit.

Pluto's moon, Charon, is about one-half the size of the parent planet. Charon is thus larger in comparison to Pluto than any other moon-planet combination in the Solar System.

Proper motion

Even though we talk about the fixed stars, there are no such stars. All stars, including the Sun, are in motion relative to each other. In general, stars are in motion around the center of our galaxy. (Note that all stars we see in the nighttime sky with our unaided eye are in the Milky Way, our galaxy.) The reason that we do not see the stars move relative to each other is that they are so far away. Put another way, the stars in the Big Dipper, for example, are in the same relative positions to each other as they were a few thousand years ago, as we observe them with our unaided eyes, even though their motions relative to the Sun are tens or even hundreds of kilometers per second.

By careful measurement, we can observe the motions of stars relative to the Sun. A star's motion has two components that we can measure directly: its proper motion is the angular distance that a star moves across the sky in a specified length of time, usually measured in arc seconds per year (one arc second equals 1/3,600 of a degree); and its radial velocity, or speed toward or away from the Sun measured by its Doppler shift. Well-known stars with large

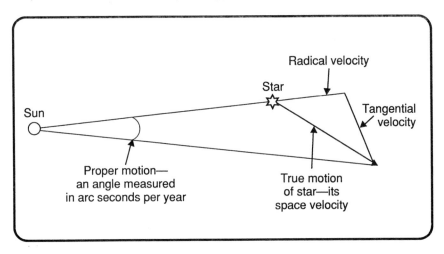

proper motions are Sirius, which moves 1.3 arc seconds per year, and Arcturus, which moves 2.3 arc seconds per year. Alpha Centauri, the second closest star to the Solar System, has the largest proper motion of any bright star at 3.7 arc seconds per year. We convert proper motion to tangential velocity simply by knowing the star's distance from the Sun, which gives us the distance the star moves in one year. Knowing the tangential velocity and the radial velocity gives us its space velocity, or speed relative to the Sun.

Calculating the space velocity of many stars near the Sun tells us how the Sun's neighboring stars are moving, which in turn gives us immediate knowledge about the motion of the stars in our region of the Galaxy relative to the galactic center. In other words, starting with proper motion, we can then interpret the rotation of our galaxy.

Ptolemy

One of the greatest astronomers, geometers, and geographers of antiquity was Claudius Ptolemy, a Greek who lived in the second century A.D. His major astronomical work is titled the *Almagest*, an Arabic word meaning *The Great Treatise*. It originally appeared in three volumes and included not only Ptolemy's astronomical work but also the work of past Greek astronomers. It thus became a detailed compilation of most of the astronomy known in Ptolemy's time. One of Ptolemy's contributions to our understanding of the heavens was his accurate calculation of the distance of our moon from the Earth. He derived a figure of 59 Earth radii (about 236 thousand miles), a number that is very close to today's accepted mean distance of about 240 thousand miles.

Ptolemy's most outstanding contribution to astronomy was a scheme for predicting the motions and arrangements of the Sun, moon, stars, and planets. This branch of astronomy is called cosmology and is concerned with the structure of all the parts of the entire Universe. (A part of cosmology that investigates the origin of the Universe is called cosmogony.) Ptolemy's cosmology was based on geometry since that was the only well-developed mathematics available to the ancients. Algebra, the calculus, and other mathematical tools had not yet been invented.

Ptolemy's Universe was geocentric; that is, he thought that the Earth was at the center of the Universe and was stationary. All the stars, the Sun and moon, and all the planets were thought to revolve around the Earth. In order to predict the motions of these objects to the accuracy of the unaided eye (the astronomical telescope was not invented until 1609), Ptolemy devised a complex scheme of circles called epicycles and deferents that were supposed to represent the orbits of the celestial bodies around the Earth (see drawing p. 118).

He further thought that, with the Earth at the center of the Universe, the other objects in order were the moon, Mercury, Venus, the Sun, Mars, Jupiter, and Saturn. (Uranus, Neptune, and Pluto can be seen only with a telescope; thus, the naked-eye planets plus the Sun and moon were the only objects, aside from the stars, known to the ancients.) The stars were thought to be not only fixed relative to each other, but were also thought to be the same distance from the Earth. They were supposed to be attached to a crystalline sphere that turned around the Earth. Beyond the sphere of the fixed stars was an additional sphere called the *primium mobile*, the "first mover" or "first cause," that made all the objects in the Universe move around the Earth.

Today, of course, we know that the Earth is not the center of the Universe. The Earth is a rather small planet among nine that revolve around a star, the Sun, that is in a collection of four hundred billion stars we call the Milky Way Galaxy. Our galaxy is only one of billions of galaxies that we can photograph with our largest telescopes.

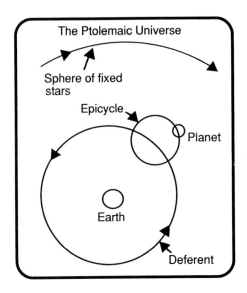

Even though Ptolemy's geocentric Universe gave way in the sixteenth century to Copernicus's heliocentric (Sun-centered) Universe, it endured for 14 hundred years, longer than any other western cosmology since before the Greeks. It is little wonder that Ptolemy has been called the most important astronomer of ancient times.

Quasars

In 1960, two star-like objects were identified as emitters of radio energy. These objects were called quasi-stellar-radio sources because they superficially resemble stars as well as emit long-wave radio energy. The name has been shortened to quasar.

The nature of these objects remained unknown for several years; they appeared as if they were made of no known chemical elements. In 1963, Maartin Schmidt of the Hale Observatories suggested that these objects are moving rapidly away from our galaxy and that they are therefore at great distances from us. (Remember, if we can measure the speed an object is moving away from us we can calculate its distance.)

One recently discovered quasar is moving away from us at 95 percent of the speed of light. This enormous speed indicates that the quasar may be billions of light-years distant from the Milky Way Galaxy. If we assume that the Universe is up to 20 billion years old, the amount of time that has passed since the Big Bang, then the light from this quasar has been traveling toward us for most of the history of the Universe. We are seeing this object as it appeared shortly after the creation of the Universe (or at least after the Big Bang) because when we look out into space we are inevitably looking back in time.

Put another way, we may be looking at an object very close to the limits of our observable Universe. Obviously, if an object is traveling at the speed of light away from us, the light will never reach us, and we will not be able to detect its existence.

Quasars are still some of the most puzzling objects ever discovered. If they are at the tremendous distances most astronomers think they are, they are very bright and thus are putting out prodigious amounts of energy. Because they are so far away from the Earth, however, a modest-sized telescope is required to see even the brightest quasars. Some of them are radiating in both optical and radio wavelengths, but most of them are radio quiet.

No one has yet been able to explain adequately how these objects can produce such large amounts of energy. Most quasars put out much more energy than an entire, normal galaxy. The best explanation we currently have is that quasars are the very active nuclei of galaxies and are thus made of familiar elements. The energy may be derived from supermassive black holes in cores of these galaxies. Why there are no nearby quasars has not yet been satisfactorily explained.

Not all astronomers are convinced that quasars are a long distance away. Some investigators suggest that they may be quite small, nearby, and not terribly bright. Thus, we do not need to find exotic explanations, such as supermassive black holes, to account for their brightness. On the other hand, we cannot then explain the data that indicate that these objects are moving rapidly away from us because we know of no mechanism that could accelerate nearby objects to the tremendous speeds that quasars seem to have.

Hence, quasars continue to present us with conflicting observational evidence that, despite intensive research efforts during the past 30 years, astronomers have not yet been able to explain in a satisfactory way.

Rainbows

Everyone is familiar with rainbows, those beautiful bands of color that stretch across the sky usually just before or after a rainstorm. We have also been told that there is a pot of gold at the end of the rainbow, although the correct end is never specified. People who have watched a rainbow as they drive in a car, or otherwise move in relation to the rainbow, know that the rainbow seems to maintain the same distance from the observer no matter how rapidly one approaches the rainbow.

A rainbow is a group of colored bands that are the result of light refracted and reflected from water drops in a rain shower, a waterfall, or a fountain. The drops of water are illuminated by light from the Sun or even the moon.

Let us consider what happens to sunlight as it interacts with one drop of water. The drop of water is spherical, or nearly so, and behaves as if it were a tiny prism. White light, made up of all the colors of visible light from the Sun, enters one side of the drop of water and is refracted, or bent. The blue wavelengths of light are refracted more than the green wavelengths, the green more than the yellow, the yellow more than the orange, and the red wavelengths of light are bent least of all. We say the light is dispersed into its various colors.

The refracted light hits the opposite side of the drop of water and is reflected back to the front surface. When the light emerges from the same side as it enters, it is bent again and spread into what we see as the rainbow of colors. Of course, we are really seeing the spectrum of the Sun. When this same process takes place in countless drops of water in a rain shower, we have a rainbow in the sky. Since the light emerges from the raindrops on the same side as it entered, we always see a rainbow in the opposite direction from the source of light.

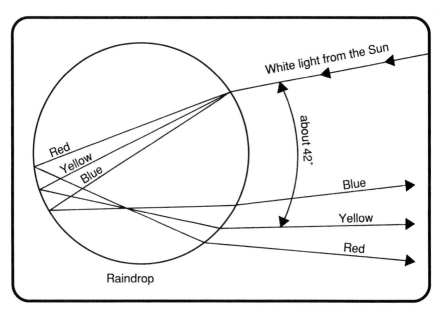

For a rainbow to be visible, the Sun must always be less than 42 degrees above the horizon because the angle between the ray of entering white light and the emerging spectrum of colors is 42 degrees. Were the Sun higher in the sky, the rainbow effect would be lost to an observer on the ground. That is why, in the summer when there are isolated showers, rainbows appear in the morning, the latter part of the afternoon, or in the evening but never around noon when the Sun is more than 42 degrees above the horizon.

The rainbow will appear in the sky as an arc that has a radius of 42 degrees from its circumference. Furthermore, the center of the arc will be exactly opposite the Sun. Obviously, since the Sun has to be above the horizon to provide light that will be refracted and reflected by the raindrops, the center of the rainbow will be below the horizon. Consequently, the closer to the horizon the Sun is when a rainbow is formed, the higher and larger the arcs will appear.

Red giants and white dwarfs

Depending on the mass it starts with, a star is born with varying characteristics such as size, brightness, color, rate of energy production, temperature, and density. As stars mature, most of them settle down to long, stable middle ages after sometimes rapid periods of development in their youths. Of course, stars have astronomically long youths, from tens of thousands of years to tens of millions of years, as well as astronomically long middle ages, from a few million to tens of billions of years.

The length of a star's youth and middle age depends on the mass it starts with, which in turn determines how rapidly its hydrogen is fused into helium and other elements. Stars with masses less than the Sun undoubtedly have very long youths and middle ages, perhaps up to many billions of years. The Sun will have a middle age of 10 billion years. Stars 10 to 80 times as massive as the Sun may have relatively short middle ages of only a few million years.

When a star has converted much of its hydrogen to other elements it collapses. Temperatures, densities, and pressures build up at the core of the star until the star explodes, greatly enlarging the outer regions of gas. The star expands, and as it does so, its pressure, temperature, and density drop in its outer regions. The star's color turns red, indicating a decrease in temperature, and the star is said to be a red giant.

Because the star's core continues to shrink after the star has reached the red giant stage, the core will attain a density up to a million times the density of water; one tablespoon of the core of this type of star, if we could put it on a scale, would weigh a thousand tons.

The Sun will almost certainly explode into a red giant star in another five billion years; what we now know as the inner regions of the Solar System will be engulfed by the expanding gases of the Sun. All the inner planets, including the Earth, will be vaporized by the 10 thousand-degree temperature of the expanding gas.

The hot core of the Sun will evolve into a white dwarf. This core will continue to shine with a surface temperature of 50 thousand degrees Kelvin. It will cool during a period of several hundred million years, eventually becoming a cinder in the Galaxy. Some astronomers have suggested that this final stage in the life of a star, which is the fate of most stars, should be called the black dwarf stage. The object is simply a cold mass of gas, much denser than any solid we are familiar with on Earth, drifting through space for the rest of the life of the Universe.

Roche Limit

A moon on the order of five hundred kilometers or more in diameter near a planet that is much more massive cannot be stable. The reason is that tidal forces caused by the gravitation of the parent planet would pull the moon apart. Put another way, the gravitational forces of the planet acting on the moon would overcome the ability of the moon to hold itself together by its own gravitation.

For example, if the Earth had a moon much closer than the present one, the gravitational pull on the side of the moon facing the Earth would be so much greater than the gravitational pull on the moon's backside that this hypothetical moon would be unstable.

In 1850, Edouard Roche showed that if a moon of the same density as the parent planet were to come within 2.44 times the radius of the planet (he used Saturn as an example), the moon would break apart. This distance is called the Roche Limit.

The Roche Limit for the Earth is 15,562 kilometers, or about 9,200 kilometers above the surface of our planet. If our moon, which has a density similar to the Earth's, were to come within 9,200 kilometers of our surface, it would shatter into countless fragments that would gradually spread into a thin ring above our equator. We should also note that the tidal disruptions at the surface of the Earth would be massive due to the moon's gravitational field, but the Earth would not break apart.

All the major moons in the Solar System are beyond their parent planets' Roche limits. A few very small moons, much less than five hundred kilometers in diameter, are within their planets' Roche limits.

It should not come as a surprise that all the planetary ring systems—around Jupiter, Saturn, Uranus, and Neptune—are inside the Roche Limit for each of these planets. The gravitational fields of these four planets simply will not allow the particles that make up the rings to accrete into one or more large moons.

It is not yet clear in our understanding of the dynamics of planetary rings whether the particles came first and never were permitted to accrete into a moon, or whether one or more moons wandered into the Roche Limit for their planet and were shattered by the tidal forces.

Insignificant masses such as scientific satellites in near-Earth orbits are not subject to Roche Limit constraints. They are simply too small for tidal forces to be effective. Further, they are held together by their tensile strengths rather than by self-gravitation. Even large satellites such as the proposed space station would be unaffected.

Saturn

Saturn, the Roman god of agriculture and the father of Jupiter in ancient mythology, is the sixth planet from the Sun. It is also the farthest planet from the Sun (at a distance of more than 1.4 billion kilometers) that can be seen without a telescope. It travels around the Sun in about 29½ Earth years.

The planet is 95 times the mass of the Earth and has a diameter of 115 thousand kilometers. Its density is lowest of any of the planets in the Solar System, having a value of only 7/10 the density of water. Thus, if we had a bathtub large enough, we could put Saturn in it and the planet would float.

Saturn's structure is undoubtedly similar to that of Jupiter. It is a large ball of gas, primarily hydrogen, with perhaps a rocky core. Saturn rotates on its axis in little more than 10½ hours, meaning that its equatorial speed is on the order of 36,000 kilometers per hour. The Earth's speed of rotation at its equator is a mere 1,667 kilometers per hour. Due to its high rotational speed, Saturn is oblate; that is, it is flattened at its poles and bulges at its equator.

The temperature at the top of Saturn's atmosphere has been measured at 95 degrees Kelvin (288 degrees below zero Fahrenheit). Similar to Jupiter, Saturn puts out more energy, particularly at the longer wavelengths, than it receives from the Sun. However, when we observe Saturn we are seeing only reflected sunlight. In addition to hydrogen and helium, both methane and ammonia have been detected in the planet's upper atmosphere.

Galileo saw Saturn's rings with the first astronomical telescope at the beginning of the seventeenth century. However, he was not able to determine precisely what he was observing. It was not until 1659 that the ring structure was identified by the Dutch astronomer, Christian Huygens. From the Earth, even a small telescope reveals three rings around Saturn. Voyager 1 and 2 pictures resolved these rings into a dazzling display of hundreds, if not thousands, of ringlets.

The rings cannot be solid because Saturn's gravitational field would disrupt any solid structure that close to the planet. Data analyzed from the Voyager missions suggest that the rings are made of countless chunks of water ice and rock covered with water ice. These chunks are a few centimeters to several meters in diameter. Sunlight is reflected from these particles, giving the illusion from the Earth that Saturn is surrounded by solid rings. When the rings are edge-on to our line of sight, they are not visible, indicating that they are no more than a few kilometers thick.

No one is certain, of course, how these icy particles all came to orbit the planet in very nearly the same plane. The best theory is that the parent planet's gravitational field is large enough to keep debris left over from the process of formation of the planet from accreting into another moon.

Saturn has 20 known moons. The largest of these satellites is Titan, a moon larger than ours and one of only two moons in the Solar System known to

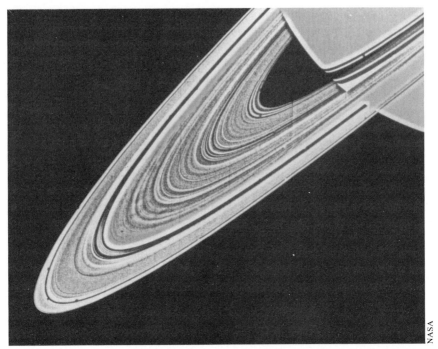

Voyager 1 image of a part of Saturn and its rings from eight million kilometers (five million miles). The close-up photo shows that the ring system we see through Earth-based telescopes is made up of hundreds of thin ringlets.)

have an appreciable, permanent atmosphere. Nitrogen is the principal gas in the atmosphere, which also has trace amounts of methane and other hydrocarbons.

The seasons

If we make careful note of the place where the Sun rises in the east each morning in the late winter (provided we get up early enough and the sky is clear) or where the Sun sets each evening in the west, we will see that it rises and sets farther north each day. The Sun appears to move during the course of a day across the sky from east to west in the southern sky (as seen from the Northern Hemisphere). The farther north it rises each day, the farther it has to go to cross the sky and the longer the periods of daylight become each day. Finally, on the first day of spring, about March 21, the Sun rises directly in the east (much to the bother of people driving east to work) and sets directly in the west (much to the bother of the same people returning home from work).

Each day after the first day of spring the Sun continues to rise and set farther north of east and west until on the first day of summer, about June 22, the Sun rises and sets about 23½ degrees north of east and west, respectively. This is as far north as the Sun goes. The first day of summer is, of course, the longest day of the year because the Sun has to travel the greatest distance across the sky.

Thereafter, the Sun begins moving south again until on the first day of autumn, about September 23, it rises and sets directly east and west again. The Sun continues to rise and to set farther south until the first day of winter, around December 23, when it is as far south as it ever moves. It then starts its journey back north once again.

On the first day of winter, called the winter solstice (meaning stationary Sun), the Sun at noon, when it is at its highest in the sky, can never be more than about 26½ degrees above the southern horizon in Illinois, for example. No wonder this is the shortest day of the year; daylight lasts little more than 9 hours in this part of the world. Conversely, the Sun is up for about 15 hours on the longest day of the year (the summer solstice); it is 73½ degrees above the horizon at noon on this day in Illinois. At the vernal equinox (the first day of spring) and the autumnal equinox (the first day of autumn), the Sun is 50 degrees above our southern horizon at noon in Illinois, and we have about 12 hours of daylight and 12 hours of darkness.

The reason for the changes in the altitude of the Sun—and consequently for the changes in the seasons—is that the Earth's axis always points almost directly at Polaris, the North Star. This situation means that the Earth's equator is tilted 23½ degrees to the plane of its orbit. Consequently, the Sun shines more directly on the Earth's Northern Hemisphere from about March 21 to September 23, when we see the Sun rise and set north of east and west, resulting in our spring and summer. It shines more directly on the Southern Hemisphere from September 23 to March 21, when we see it rise and set south of east and west, resulting in spring and summer south of the equator and fall and winter north of the equator.

It is also true that the Earth, because it goes around the Sun in an oval (elliptical) orbit, is closest to the Sun in January by several million kilometers. This circumstance means that summer in the Southern Hemisphere should be warmer than summer in the Northern Hemisphere and that winter in the Northern Hemisphere should be milder than winter in the Southern Hemisphere. It turns out, however, that there is more ocean in the Southern Hemisphere; oceans tend to have a moderating effect on climate, and thus the winters and summers in the two hemispheres are rather equal (see drawing p. 126).

We know that there is a lag in the seasons. Even though the Northern Hemisphere receives the least light from the Sun on December 22, the coldest days of winter are usually several weeks later. The reason is that the land and water masses are still cooling from the summer. In summer, the hottest day is not June 22 but occurs toward the end of July. In June, the land and water masses are still warming from the winter.

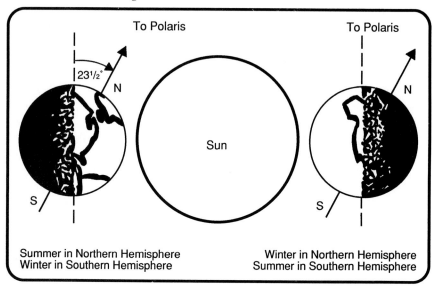

Summer in Northern Hemisphere
Winter in Southern Hemisphere

Winter in Northern Hemisphere
Summer in Southern Hemisphere

The Solar System: An overview

Our small part of the Milky Way Galaxy is called the Solar System. It contains a primary star (the Sun), nine planets, 63 planetary moons, several thousand known minor planets (asteroids), meteoroids, comets, interplanetary gas and dust, and charged particles from the Sun called the solar wind. The Solar System is held together, of course, by the gravitational field of the Sun.

In comparison to other nearby stars, the Sun is somewhat greater than average in mass, size, and luminosity (brightness). About 99.86 percent of the mass of the Solar System is contained in the Sun. The total mass of all the planets is only about .14 percent of the entire mass of the Solar System; the moons in the system make up only .00005 percent of the total mass. Comets, meteoroids, the minor planets, and the interplanetary gas and dust account for little more than .01 percent of the total mass of the Solar System. Thus, we can say that the Sun, at least in terms of mass, is the Solar System.

The four inner planets—Mercury, Venus, Earth, and Mars—are often called the terrestrial planets because they are, in a general way, composed of the same materials: rock and metal. They are all also nearly the same size (ranging from 4,878 kilometers in diameter for Mercury to 12,756 kilometers in diameter for the Earth) and are cool compared to the Sun. Of the terrestrial planets, only Mercury does not have an atmosphere. The next four planets, in order from the Sun, are Jupiter, Saturn, Uranus, and Neptune. They are often referred to as the Jovian planets because they are all large and because they resemble Jupiter in composition: they are made of gases, mostly hydrogen. Pluto is the farthest planet from the Sun that we know of and seems to be made of frozen water, some rock, and gases.

All the planets revolve around the Sun from west to east (counter-clockwise as seen from the north), and most of them rotate on their axes in the same direction. The only exceptions are Venus, which rotates slowly from east to west, and Uranus, which has its axis of rotation close to the plane of its orbit.

The orbits of the planets are nearly in the same plane. Only the plane of Pluto's orbit is inclined more than a few degrees from the plane of the orbits of the other planets.

The age of the oldest moon rocks and of the meteorites is about 4.5 billion years. Other measurements indicate that the Sun is about the same age. Somewhat less than 5 billion years ago the Solar System condensed from a cloud of gas—mostly hydrogen and helium, dust, and other material as a result of mutual gravitational attraction between all the atoms and molecules in the cloud of material. The original cloud, of course, was much larger than the present Solar System.

As the cloud contracted, the inner, denser part formed the proto-Sun. When the temperature of the interior of the proto-Sun reached about one million degrees Kelvin, nuclear reactions began to occur (hydrogen atoms fused into helium atoms), energy was released in the process, and our Sun was born as it began to shine by its own light about 4.6 billion years ago.

As the cloud of material continued to contract, all parts of it rotated faster and faster, much as an ice skater rotates more rapidly as she draws her arms closer to her body. Matter at some distance from the proto-Sun began to move fast enough to keep from falling farther into the Sun. The material that was left behind accreted over a long period of time into the various planets, moons, and other objects of the Solar System with which we are familiar today.

The Solar System: Motions of the planets

The word planet comes from the Greek word for wanderer. As seen from the planet Earth, the other planets in the Solar System seem to wander against the background of stars in a narrow band, called the Zodiac, that stretches 360 degrees around the sky.

Mercury and Venus are called inferior planets because they are closer to the Sun than the Earth and therefore have smaller orbits. Mars, Jupiter, Saturn, Uranus, Neptune, and Pluto are known as superior planets because they are farther from the Sun than the Earth and thus have larger orbits. Only Mercury, Venus, Mars, Jupiter, and Saturn can be seen without a telescope.

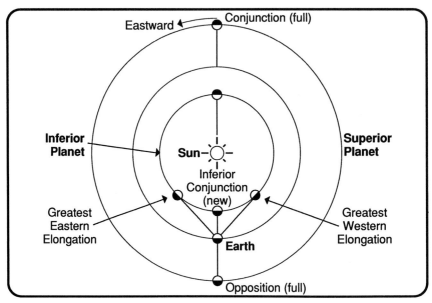

The Solar System as seen from the North.

The maximum angle between the Earth and an inferior planet (Mercury and Venus) is called greatest eastern (or western) elongation. When Mercury and Venus are at greatest eastern elongation, they are east of the Sun and therefore visible in our western sky after sunset. When they are at greatest western elongation, they are west of the Sun and are visible in the eastern sky before sunrise. Mercury's greatest elongation is 28 degrees. This small angle means that Mercury, as seen from the Earth, is never very far from the Sun; that is, Mercury can only be seen in morning or evening twilight and is, therefore, always difficult to see.

Venus's greatest elongation is 48 degrees; therefore it can appear in a dark sky. When we see Venus near its brightest in the morning or evening sky, we call it the morning star or evening star, respectively. Of course, it is a planet and not a star.

Each planet varies in brightness, as seen from the Earth, due to a combination of its distance from the Sun, its distance from the Earth, and the amount of the sunlit side visible from the Earth (in other words, its phase). When Venus is at its brightest, it is the third brightest object in the sky, behind the Sun and moon.

Superior planets have both prograde (direct) and retrograde motions against the background of stars as seen from the Earth.

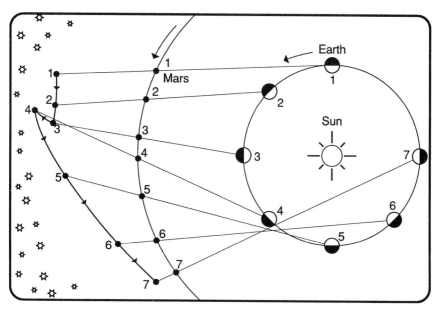

Retrograde motion of a superior planet.

All planets also have two orbital periods. The first of these is called the sidereal period, which is simply the amount of time a planet takes to revolve once around the Sun with respect to the stars. (The word sidereal is one of the Latin words for star.)

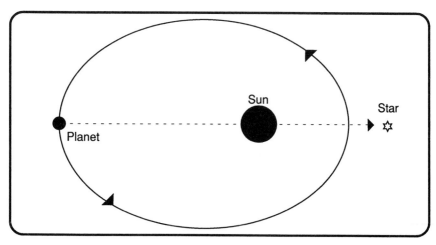

Sidereal period of any planet.

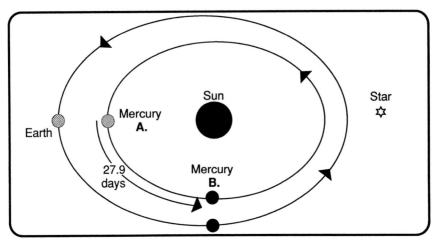

Synodic period of an inferior planet (Mercury, for example).

   A. *At inferior conjunction. Returns to this position relative to Sun and a star in 88 days—called the sidereal period.*

   B. *Mercury requires 115.9 days (88 days   27.9 days) to go from one inferior conjunction to the next inferior conjunction—the synodic period.*

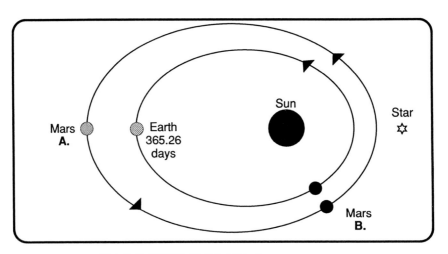

Synodic period of a superior planet (Mars, for example).

   A. *Mars at opposition. Returns to this position relative to the Sun and a star in 687 days—the sidereal period.*

   B. *Mars requires 780 days to go from one opposition to the next opposition—the synodic period.*

The synodic period is the amount of time a planet takes to return to the same arrangement of the Sun, the Earth, and the planet as seen from the Earth. For the inferior planets (Mercury and Venus), this is the time it takes for them to move between successive inferior conjunctions, for example. For the superior planets (Mars, Jupiter, Saturn, Uranus, Neptune, and Pluto), this could be the amount of time between successive oppositions (when a superior planet is on the opposite side of the Earth from the Sun).

Planet	Sidereal Period	Synodic Period
Mercury	88 days	116 days
Venus	225 days	584 days
Earth	365 days	- - - - - - -
Mars	687 days	780 days
Jupiter	12 years	399 days
Saturn	29.5 years	378 days
Uranus	84 years	370 days
Neptune	165 years	368 days
Pluto	249 years	367 days

Space: Contents

Space between the planets, space between the stars, and space between the galaxies appears to be empty. Among the planets, however, there exists gas and dust that are known as the interplanetary medium. Interplanetary dust is made up of micrometeorites, tiny particles of rock in orbit around the Sun. On dark, clear nights this material reflects light from the Sun, which is on the other side of the Earth. This band of light, reflected from interplanetary dust, is called the zodiacal light because it follows the zodiac, that region of the sky through which the Sun and most of the planets appear to move.

Interplanetary gas is composed of charged particles—protons, electrons, and ions—flowing from the Sun. This material, called the solar wind, streams out from the Sun through the entire Solar System.

The total amount of interplanetary dust and gas is quite small. Its distribution is such that, near the Earth, there are only a few particles per cubic centimeter, thus providing a better vacuum than can be produced in a laboratory.

Space between the stars also contains an interstellar medium, again composed of gas and very small solid particles. The gas and dust between the stars are not distributed uniformly, but rather tend to collect in nebulae (the Latin word for clouds). It is out of these clouds of gas, mostly hydrogen with some helium, and dust that new stars are formed.

In our part of the Milky Way Galaxy it is estimated that there is one atom of gas per cubic centimeter and 25 to 50 particles of solid material, each less than 1/10,000 of a centimeter in diameter, per cubic kilometer. Again, these densities are far less than the best laboratory vacuums.

It is also thought that gas and dust probably exist between galaxies. Faint luminosities that do not seem to be associated with any specific galaxy, as well as luminosity that appears to connect some pairs of galaxies, have been photographed through telescopes.

Space: Einstein's lesson

We think we know what space is. It is what astronomers study and astronauts explore; it is where the planets, stars, and galaxies are. These definitions of space are all correct, but we can be much more precise in our understanding of space.

Many astronomers, including the Greeks, Newton, and astronomers to this century, thought that space was absolute; that is, space exists in three dimensions—height, width, and depth—that are independent of the objects that are in it. Furthermore, we usually think of space as being bounded as the walls of the room in which you are reading this page define the space of the room. In the same way, we could think of the Universe as having some sort of boundaries enclosing space. We could even think of the Universe as having an infinite amount of space, and therefore think of it as unbounded, and still consider space as absolute.

Most astronomers are reasonably certain that the Universe is expanding, i.e., galaxies are moving away from one another throughout the Universe. If this is true and if we could go to one of the galaxies on the edge of the Universe, could we not poke our finger through the edge (or boundary) of the Universe? What would be on the other side? We are thinking of the Universe as if it were a big room; if we could move to one wall of the room and poke our finger through the wall, we could reach into the other side.

It turns out, however, that the Universe is not like a big room. Albert Einstein showed us that space is not absolute but that the characteristics of space are dependent on the mass of the objects in it. The mass of the Universe is expanding, causing space to expand with the mass. Put another way, there is nothing outside the space of the Universe. Space expands but it is not expanding into anything; the Universe increases in size only because the expanding mass creates space that also expands.

If all of this sounds strange, it is only because it is not part of our normal, everyday experience. On the other hand, it is really our every day experience—such as sticking our finger through a hole in a wall to see what is on the other side—that is rather restricted and peculiar compared to the normal events of the Universe as a whole.

We have also heard about space warps. A space warp is simply the curvature of space in the vicinity of a massive object in the Universe. Even though the Sun is not very massive compared to an entire galaxy, it is massive enough to curve or warp space around it to an extent that can be measured. When we measure a ray of light from a star that passes next to the Sun, we discover that the ray is deflected almost two arc seconds from the path it would travel were the Sun not there.

We can see starlight near the Sun only during a total solar eclipse. By carefully measuring the position of stars at night and then comparing the positions of those same stars whose light must pass close to the Sun during a total solar eclipse, the curvature of space created by the Sun can be verified, within the limits of observational techniques, as predicted by Einstein. The experiment to prove the curvature of space around the Sun was first carried out during the solar eclipse of 1919 and repeated during many eclipses thereafter.

Extending our thinking from the slight curvature of space around the not-very-massive Sun to the mass of the whole Universe, it is easy to see why astronomers say the Universe is curved. The entire mass of the Universe creates a curvature of space, and we are correct when we refer to the Universe as existing in curved space that is defined by the total mass of the Universe. Thus, the old notion of absolute space is false. Space exists only where matter exists, and its curvature is dependent on the amount of matter present.

The spectrum

All radiative energy—not only visible light but also gamma rays, x-rays, ultraviolet, infrared, and radio energy—is made up both of particles, called photons, and waves. We have all had the experience of putting a glass prism into sunlight and making a rainbow of colors, a spectrum of sunlight. Each color in the spectrum is the result of a different wavelength. In visible light, red has long wavelengths and violet has short wavelengths. As we might expect from examining a spectrum, orange wavelengths are a little shorter

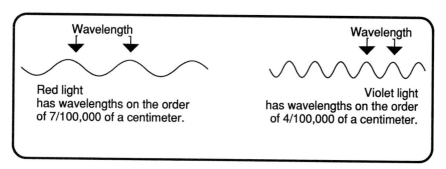

Wavelength

Red light
has wavelengths on the order
of 7/100,000 of a centimeter.

Wavelength

Violet light
has wavelengths on the order
of 4/100,000 of a centimeter.

than red but longer than yellow, yellow wavelengths are longer than green, green are longer than blue, blue longer than indigo, and violet are the shortest of all. The sequence of colors in the visible part of the spectrum can be remembered by saying ROY G BIV.

Outside of the visible light part of the spectrum, gamma rays have the shortest wavelengths—less than 1/100,000,000,000 of a centimeter—of any radiative energy. Successively longer wavelengths are found in x-rays, ultraviolet, visible, infrared, and radio waves. The latter have wavelengths up to several kilometers.

There are three kinds of spectra. A continuous spectrum is produced by a hot metal, liquid, or dense gas. An emission, or bright line, spectrum is produced by a hot gas that is less dense. An absorption spectrum is produced by a cooler gas. Almost all stars have a continuous spectrum, produced by the hot, dense gas of the star, on which dark lines are superimposed because some of the energy from the surface is absorbed by the cooler, less dense gas of the atmosphere of the star.

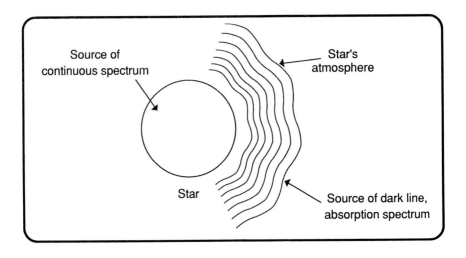

There will be an occasional star that is hot enough to superimpose bright emission lines on the continuous spectrum together with the dark absorption lines.

A spectrograph is put on a telescope to divide the light of a star, galaxy, or planet into its wavelengths. A spectrograph functions in much the same way that a glass prism works in sunlight. We obtain a spectrum of the various wavelengths of light from the object we are studying. Lines appear in the spectrum because we have introduced a narrow slit in the path of light.

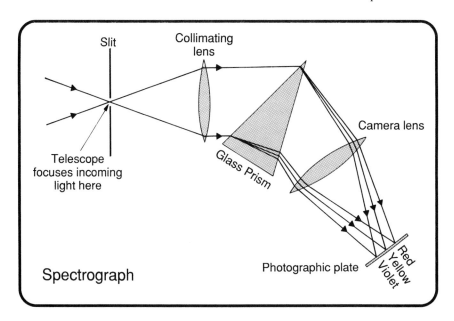

Each of the 92 naturally occurring elements in the Universe, from hydrogen and helium (the two most abundant elements) through uranium, produces or absorbs wavelengths of light characteristic of that element and its temperature (energy state). By measuring the wavelengths of light from a star, for example, we can determine the composition of the star. We can also

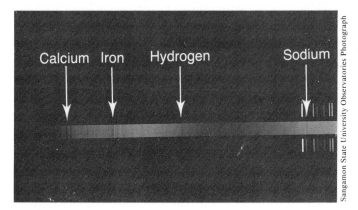

Spectrum of the Sun. Note dark absorption lines of the elements, calcium, iron, hydrogen, and sodium.

determine its temperature. More careful analysis of the spectrum will reveal the relative abundance of the elements in the star and whether the star is moving toward us or away from us and how fast (the Doppler effect). We can also learn much about turbulence in the star's atmosphere as well as how fast, if at all, the star is rotating. The spectrum also tells us the strength of the star's magnetic field.

When we analyze the spectrum of a galaxy, we are analyzing the combined light from all of its stars. Spectra of the planets give us the Sun's spectrum, because the planets all reflect sunlight, plus spectral information about the planets' atmospheres (all planets except Mercury have an atmosphere). Our moon, of course, gives us only the Sun's spectrum, because the moon does not have an atmosphere. We have learned most of what we know about stars and galaxies as a result of studying their spectra.

The Star of Bethlehem

One of the most famous stars in the Western world is the Star of Bethlehem. Closely associated with the birth of Jesus of Nazareth, which is celebrated in December, it has also been the subject of speculation by astronomers. There are several plausible astronomical explanations for this star.

We must remember that people living nearly two thousand years ago were superstitious; they called almost any bright object in the sky a star. Furthermore, people such as the biblical shepherds spent a great deal of time outdoors in the clear, dry climate of the eastern Mediterranean. They would thus be familiar with the nighttime sky; any infrequent or unusual celestial event would be the subject of their immediate attention. It is possible, for example, that the Star of Bethlehem was a bolide, or fireball. Occasionally, large chunks of metal or rock, in orbit around the Sun, burn up in our upper atmosphere due to friction when we run into them. We then see a particularly bright meteor. It seems unlikely, however, that a bright meteor, which might be regarded as a sign or portent by superstitious people, that lasts at most only a few seconds in the sky would induce the wise men to travel across a desert for many days and nights to Bethlehem.

An alternative explanation might be a comet that, if unusually large and bright, would be visible for several weeks in the nighttime sky. In 11 B.C., Halley's comet was visible during one of its 76-year trips around the Sun. Perhaps its head and tail seemed to point in the direction of Bethlehem. We know, of course, that comets are not stars, but to the ancient peoples of the eastern Mediterranean comets were often referred to as bearded stars.

Even though 11 B.C. does not coincide with the date of the birth of Jesus from which we reckon our calendar, we must remember that the date of his birth is not known accurately: it may have been as early as 11 B.C., or it may have occurred in 6 B.C. or even 4 B.C. according to our present calendar. Any bright or unusual object in the sky during this period might have been the Star of Bethlehem.

Another explanation of the star might be the conjunction (coming together) of three planets—Jupiter, Saturn, and Mars—in February 6 B.C. in the constellation of Pisces, the Fishes. (We must remember that the planets only appear to be close together in the sky as we see them from the Earth. In reality they are separated by hundreds of millions of kilometers.) Pisces was an important constellation to the Hebrews because Moses was thought to have been born when Jupiter and Saturn appeared together in that constellation. Since this conjunction of Jupiter, Saturn, and Mars is a rare event, occurring about every eight hundred years, none of the people living at that time would have seen it previously. Thus, when the planets came together again, superstitious people would think that something important was about to happen.

There are two objections to the conjunction theory. One is that planets are not stars. Yet, to the uninformed people of two thousand years ago a conjunction of planets might be called a star. The second objection is that the conjunction took place in the western sky, and the Bible clearly says that "we have seen his star in the East." But the Bible also says that "there came wise men from the East." If they came from the east they might have been traveling toward the West, following the star, and the first passage might thus be interpreted as saying, "We in the East have seen his star."

It should also be pointed out that December 25, Christmas Day, was originally a pagan Roman holiday celebrating the birth of the unconquered Sun. The winter solstice, the shortest day of the year, occurs at this time when the Sun is lowest in the southern sky. It begins to move higher in the sky with each passing day after about December 21, thus promising the return of spring.

Finally, the Star of Bethlehem might have been a nova (literally, a new star), a name astronomers give to existing stars that suddenly explode and become hundreds or thousands of times brighter for a few days or weeks. They then gradually fade from view. Some novae, called supernovae, are so bright that they are slightly visible during the day. Since people in those days knew the sky intimately, any nova would have been immediately noticed and regarded as a portent of a significant event.

Although the details are not precisely known, astronomers do know a great deal about the mechanisms that make stars explode. There may be as many as two or three dozen nova events in our galaxy each year, but nearly all of them are so far away that they are too faint or too obscured by intervening clouds of interstellar dust to be seen either with the unaided eye or with a telescope. Only infrequently are novae able to be seen with no optical aid.

Thus, we have several astronomical explanations for the Star of Bethlehem. Among these you may choose the one that you like the best, or you may prefer to believe in a nonscientific explanation.

Star brightness and distance

Stars appear to our eyes with differing brightnesses for two interrelated reasons. First, stars have intrinsically different brightnesses: some are much brighter than others because they are hotter, thus emitting more energy, or they are simply larger. Second, stars are at varying distances from the Earth: faint stars nearby may appear brighter than luminous stars that are farther away.

Hipparchus, the greatest astronomer of the pre-Christian era, divided stars by their apparent brightnesses, or the way they appear in the sky, into six magnitudes. Each star in his catalog was assigned a magnitude: the brightest stars were first magnitude and the faintest stars were sixth magnitude. In his scheme, which we still use in modified form today, the smaller the number, the brighter the star; obviously, the larger the number, the fainter the star.

The second brightest star in the sky is Sirius, the Dog Star, with an apparent magnitude of about -1.5. The brightest star in the sky is the Sun, with an apparent magnitude of nearly -27. There are thousands of sixth-magnitude stars, those that are just visible to the best human eyes under a perfectly clear and dark sky.

If we could somehow move all the stars to the same distance from the Earth, we would eliminate the differences of brightness among them as a result of distance. Stars would then differ only because they are intrinsically different in brightness. Of course, we cannot really move stars around in the sky, but astronomers can mathematically put all stars at the same distance and then compare their absolute magnitudes.

Astronomers have arbitrarily chosen 10 parsecs (32.6 light-years) as the distance at which intrinsic brightnesses of all stars should be compared. At this distance, the Sun would barely be visible to the unaided eye with an absolute magnitude of +4.8. Deneb, a star in the constellation of Cygnus, the Swan, has an apparent magnitude of just more than +1. If it were at 10 parsecs it would have an absolute magnitude of -7, and would be one of the brightest stars in the sky. Obviously, its real distance is more than 10 parsecs; indeed, it is 1,500 light-years from the Earth. Polaris, the North Star, would also be considerably brighter than it apparently is at a distance of 820 light-years.

In absolute magnitudes, the brightest stars are more than 60,000 times brighter than the Sun, and the faintest stars are about 1/10,000 as bright as the Sun. The two hundred-inch telescope on Palomar Mountain, using sensitive electronic detectors, can just detect stars with apparent magnitudes of about +26, which is 4,000,000,000,000,000,000,000 times fainter than the apparent magnitude of the Sun.

Of course, other astronomical objects such as galaxies, star clusters, planets, and the moon are also classified in terms of their brightnesses on the same magnitude scale. The moon has a maximum apparent magnitude, when it is in its full phase, of -12.5. Venus has a maximum magnitude of about -4 and is then the third brightest object in the sky after the Sun and moon.

Star names: Winter sky

Most of the bright stars have names that are of Greek, Latin, or Arabic origin. Some of the best-known bright stars that can be seen in the winter sky include the following.

Capella: In the constellation of Auriga, the Charioteer, this bright star has a name derived from Latin that means little she-goat. The Charioteer is often pictured with a goat in his left arm.

Sirius: Pronounce the name of this star as if it were spelled serious. This is the second brightest star in the sky (the Sun is the brightest); its name is from the Greek, meaning the scorching one. It is in the constellation of the Big Dog (Canis Major) and is also called the Dog Star.

Procyon (Prō' sĭ-ŏn): This term comes from the Greek, meaning before the dog. This star, in Canis Minor (the Little Dog), is a bright star that rises before Sirius.

Castor and Pollux: These are the Latin forms of the Greek names for the twin sons of Zeus. These two stars of nearly the same brightness are in the constellation of Gemini, the Twins.

Regulus: The name is the Latin diminutive of rex, the king, and thus means prince. It is the brightest star in the constellation of Leo, the Lion, king of beasts.

Betelgeuse (Bĕt'ĕl-jūz): The name is probably from the Arabic word that means armpit of the white-belted sheep. This very bright star was originally in a group that the Arabs saw as a sheep. It still retains its meaning because it represents that armpit or shoulder of Orion, the Great Hunter.

Rigel (Rī'jĕl): The second brightest star in Orion, the word is from the Arabic and means Orion's left foot.

Aldebaran (Ăl-dĕb'ă-răn): This name is also from the Arabic language and means the follower. The star follows the Pleiades across the sky and is the brightest star in the constellation of Taurus, the Bull. Aldebaran is located in the bull's face and is often pictured, because the star is red, as the bull's eye. The group of stars making up Taurus's face is called the Hyades, known in mythology as the daughters of Atlas and Aethra.

Star names: Summer sky

The following are some of the prominent bright stars in the summer sky.

Altair (Ăl'târ): A star that forms one of the corners of the summer triangle, Altair is an Arabic word meaning the flying eagle or vulture. It is the brightest star in the smaller constellation of Aquila, the Eagle.

Deneb (Děn'ĕb): This star is also one of the corners of the summer triangle. Its name comes from the Arabic and means tail of the hen. It represents the tail of Cygnus, the Swan. Deneb is one of the brightest stars we know of; if it were located little more than 30 light-years from the Earth, it would be brighter than our moon is at first quarter. Its real distance is 1,500 light-years.

Vega (Vē'gȧ): The third star in the summer triangle, Vega is also an Arabic word and means the falling eagle or vulture. It is in the small constellation of Lyra, the Lyre, or harp.

Arcturus (Ȧrk-tū-rŭs): The name of this star is from the Greek and means the bear guard. It is in the constellation of Bootes, the Bear Herder who drives the Big Bear (Ursa Major) around the sky. It is the third brightest star, behind the Sun and Sirius (the Dog Star), that we can see from the central part of the United States.

Antares (Ăn-tā'rēz): It is a beautiful, red star in the constellation Scorpius, the Scorpion. The star's name is from the Greek, meaning the rival of Mars. Mars, too, is a red object in the nighttime sky. Antares is often thought of as the tail of the scorpion.

Spica (Spī'kȧ): This is a Latin word meaning ear of corn, but is sometimes translated as sheaf of wheat. Spica is the most conspicuous star in the constellation of Virgo, the Virgin, or young woman.

Polaris: Although there are 50 stars brighter than Polaris, it is included in this discussion of prominent stars because it is the only star in the sky that never seems to move. Because the Earth's North Pole points within one degree of Polaris, it is always in the same place in the northern sky, giving a nearly true indication of the direction of north. If we can measure its altitude above the northern horizon, Polaris also gives us our latitude. It thus is an invaluable aid to navigators on the open ocean. Due to the Earth's rotation, all the other stars seem to circle around Polaris. The word Polaris is from the Latin phrase *stella polaris*, meaning the north star.

Stars

A star is defined as a sphere in the sky that gives off its own light. The least massive stars are thought to have a mass only about one-tenth the mass of the Sun. Any mass less than this would not have enough gravity to create the necessary temperature and pressure in the object's core to trigger nuclear fusion (hydrogen to helium), the source of light.

Jupiter, the most massive object in the Solar System after the Sun, has only about 1/1,000 the mass of the Sun. It thus would need to be about 100 times more massive than it is to classify as a star by fusing hydrogen to helium (note that the Earth is 318 times less massive than Jupiter). The most massive stars may be on the order of 80 times the mass of the Sun. Any object more massive is thought to be unstable.

The coolest stars have surface temperatures less than 3 thousand degrees Kelvin, or about twice as hot as melted steel in a furnace. The hottest stars have surface temperatures in excess of 50 thousand degrees Kelvin (100 thousand degrees Fahrenheit). These temperature ranges are for stars that are converting hydrogen to helium in their cores, about 90 percent of the nearby half million or so stars in the solar neighborhood. The other 10 percent of the stars produce energy, and hence light, by other fusion processes in their cores, such as helium fusing to carbon and carbon fusing into heavier elements up to iron.

Most stars are about 80 percent hydrogen and 20 percent helium. All the other 90 naturally occurring elements together make up from a tiny fraction of one percent of the mass of a star to about two percent of the mass of a star. By studying these tiny differences in the abundances of elements heavier than hydrogen and helium, astronomers are able to form a general picture of the chronological sequences of stars. (To the astronomer, all elements heavier than hydrogen and helium are called metals. This term is not used in the chemical sense, but rather as a way of indicating any element in a star that is not hydrogen or helium.)

It was realized about 50 years ago that, in general, stars in a galaxy, including the Milky Way, can be classified in two ways. Population I stars are those stars that have relatively high abundances of metals (about two percent) and are found in the disk of a galaxy, particularly in the arms of spiral galaxies, and in denser regions of gas in a spiral galaxy. Population II stars are older stars that have relatively low abundances of metals (less than one percent) and are generally found in the cores (bulges) of spiral galaxies, in elliptical galaxies, and in globular clusters. The Sun is a population I star; that is, it is comparatively rich in metals.

There are stars with intrinsic brightnesses much greater than the brightness of the Sun. Deneb, for example, the tail of the swan in the constellation called Cygnus, is more than 63 thousand times brighter than the Sun. Sirius, the Dog Star, is only about 30 times brighter than the Sun. It appears much brighter in the nighttime sky than Deneb because it is so much closer, about nine light-years compared to 15 hundred light-years for Deneb.

The largest stars may be many times the size of the Sun. Antares, in the constellation of Scorpius, has a radius five hundred times that of the Sun. This means that if Antares were in the middle of our Solar System, its edge would be beyond the orbit of Mars.

The closest star to the Sun is Proxima Centauri (or, more accurately, the Alpha Centauri system of three stars) at a distance of about 4.3 light-years (25 million, million miles). At the speed that the Voyager 1 spacecraft is leaving the Solar System, about 60 thousand kilometers (37 thousand miles) per hour, it would take about 80 thousand years to reach this closest star to the Sun.

Stars: Classification

In 1911, the Danish astronomer Ejnar Hertzsprung discovered a relationship between the temperatures and intrinsic brightnesses (absolute magnitudes) of stars. This same relationship was also discovered independently by the American astronomer, Henry Norris Russell, in 1913. The basic relationship is that, in general, the hotter the star the brighter it is (i.e., the more energy it produces). We can develop a Hertzsprung-Russell, or H-R, Diagram for all stars.

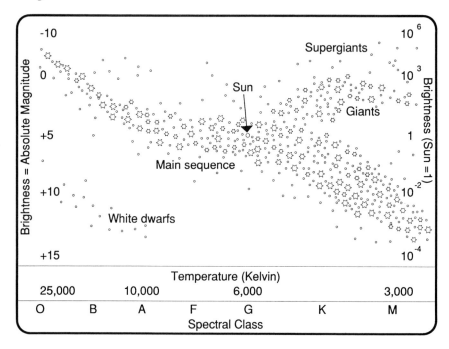

About 90 percent of the stars in the Sun's neighborhood are on the main sequence. These are stars that are fusing hydrogen to helium in their cores. Note that the Sun has an absolute magnitude of +4.8 and a surface temperature of about six thousand degrees Kelvin.

Giants and supergiants are very bright but not very hot. This means that they must have extended photospheres (they are very large), which in turn indicates that any given area on the star's surface is relatively cool, only a few thousand degrees, but that there is such a large surface that the star is very bright despite its relatively cool temperature.

The giants and supergiants are stars near the end of their lifetimes. They are puffed up because nuclear fusion processes, other than hydrogen to helium, at very high core temperatures are creating so much energy that the outer layers of these stars swell out. Hence, these stars are quite large.

The white dwarfs are very hot but small, and thus quite faint. That is, each area of their surfaces is hot, but there is not much total surface area to these Earth-sized stars. These are stars in which no fusion is taking place; they are stars at the end of their lives.

Note also that the stars in the H-R Diagram are also classified by spectral class: OBAFGKM. O-type stars on the main sequence are the hottest and brightest; the M-type stars on the main sequence are the coolest and faintest. All stars on the main sequence are referred to as dwarfs. Therefore, the Sun is a G-type dwarf. There are M-type giants and supergiants, as well as dwarfs.

It is easy to remember the spectral sequence on classification of stars by the following mnemonic: <u>O</u>h, <u>B</u>e <u>A</u> <u>F</u>ine <u>G</u>uy (<u>G</u>irl), <u>K</u>iss <u>M</u>e.

The H-R Diagram is the most useful diagram we have in astronomy. The basic diagram, as shown above, and its literally hundreds of adaptations, tell us the story of the stars.

Stars: Variables

There are several different kinds of variable stars, stars that vary in brightness for a variety of causes. For example, there are binary stars, two stars orbiting around a common center of mass. If the plane of their orbit is in or close to our line of sight here on the Earth, these stars will eclipse each other and their combined light will vary cyclically over time. About half the stars in the Sun's neighborhood are binary star systems, although few are eclipsing binaries because the plane of their orbit is not in or near our line of sight.

There are also several different kinds of intrinsic variable stars. That is, they vary in brightness due to some internal mechanism that causes them to change brightness. Pulsating variables change brightness on a regular schedule. For example, Cepheid variables, useful in determining distances to external galaxies, change about one magnitude in 1 to 50 days. RR Lyrae stars change about one magnitude in 1 to 24 hours. Long-period variables, such as Mira, change up to three magnitudes in 130 to 500 days. A few stars called dwarf Cepheids change two magnitudes in 1 to 3 hours.

The reason for these changes is that these stars are unstable. Temperatures and pressures build in the core as the star shrinks. Enough energy is created to reverse the inward movement of the outer regions of the star. The star swells and, as it does so, core temperatures and pressures drop, less energy is produced, and the star starts to shrink again, starting the cycle over.

Nonpulsating variables are stars that change brightness only once (or at least once within the modern astronomical time period). T Tauri stars, for example, are pre-main-sequence stars that are blowing off their outer shells of gas and dust, thus causing changes in brightness.

Wolf-Rayet stars are old, massive objects that are unstable and are losing mass (blowing off their outer layers) and thus also changing brightness. Planetary nebulae are also old, unstable stars that are rapidly losing mass.

Novae and supernovae are stars at the end of their lifetimes that are cataclysmic variables. They suddenly become much brighter, particularly the supernovae. Supernovae are the most energetic explosions in the Universe and probably account for all the elements in the Universe heavier than iron. They may also be a source for high-energy cosmic rays.

The Sun: Physical characteristics

The closest star to the Earth is our Sun at a mean distance of about 150,000,000 kilometers (93,000,000 miles). The visible portion of the Sun has a diameter of 1,390,000 kilometers, or about 109 times the diameter of the Earth. The average density of the Sun is about half again the density of water, but, as we shall see, there are many regions of the Sun that approximate our best laboratory vacuums.

The Sun shines because in its core, where the temperature is thought to be 15 million degrees Kelvin (about 27 million degrees Fahrenheit), hydrogen atoms are fused into helium. In this process a small fraction of the mass of a hydrogen atom, about 7/10 of one percent, is converted to energy. This energy makes its way to the surface of the Sun and is radiated into the Solar System as visible light.

In order for the Sun to shine, six hundred million tons of hydrogen must be converted every second into helium. This means that four million tons of matter are converted to energy every second in the Sun. At this rate, the Sun will be able to produce energy for about 10 billion years. Since we know that the Sun is about 4.6 billion years old, it is thus halfway through its expected lifetime.

The temperature at the surface of the Sun is about 6 thousand degrees Kelvin (about 11 thousand degrees Fahrenheit). We know that no solids or liquids can exist at this temperature and that the Sun must therefore be a big ball of gas. Because the Sun is not solid it can rotate at speeds that differ with the solar latitude. For example, its speed of rotation at the solar equator is nearly 25 days. At the poles, its speed of rotation is about 35 days.

The most common element in the Sun is hydrogen; about 80 percent of the mass of the Sun is hydrogen. Helium is the second most common element. Slightly more than 98 percent of the Sun is hydrogen and helium together. (The word helium is derived from the Greek word for Sun, *helios*, because the element was discovered in the Sun before it was discovered in the Earth's atmosphere.)

The more than 60 other elements that have been found in the Sun make up only a small percentage of the mass of the Sun. Of these, oxygen, carbon, nitrogen, silicon, magnesium, neon, and iron are the most prominent—all in gaseous form. Elements that have not been discovered either do not exist in the Sun or occur in such small amounts that we cannot detect them.

The Sun appears yellow to us simply because these are some of the dominant wavelengths of light for the temperature at the surface of the Sun. Were the Sun cooler, it would be red. If it were hotter, it would be blue or white.

The most massive stars are perhaps 80 times more massive than the Sun; the least massive stars are one-tenth the mass of the Sun. The Sun also has a diameter a hundred times greater than the smallest stars, but only 1/500 the size of the largest stars.

The Sun is 10 thousand times brighter than the dimmest stars and only 1/60,000 as bright as the brightest stars. The coolest stars have surface temperatures of about 3 thousand degrees Kelvin, and the hottest stars are at 50 thousand degrees Kelvin. The Sun, as we noted, has a surface temperature of about 6 thousand degrees Kelvin.

The Sun: Structure

In order to understand the structure of the Sun, it is important that we begin by recognizing that there are no layers or distinct boundaries in the Sun. It is appropriate to think of the Sun as a mass of gas that is very dense and very hot at the center. It gradually becomes less dense and generally cooler away from the center.

The core of the Sun has a temperature of about 15 million degrees Kelvin and a density thought to be nearly 150 times that of water. Next are two regions that transmit energy, created in the core, to the outer regions of the main part of the Sun. Energy is transmitted by radiation in the inner region and by convection in the outer region. It then reaches the photosphere, that region of the Sun that we see farthest into and from which the vast majority of visible light is emitted. The photosphere has a temperature of about six thousand degrees Kelvin, a pressure only a few hundredths of the Earth's atmospheric pressure at sea level, and a density 1/10,000 of the Earth's atmospheric density.

Just above the photosphere is the chromosphere, a region two thousand to three thousand kilometers thick with an average temperature of 15 thousand degrees Kelvin. At the center of this region the temperature is 100 thousand degrees Kelvin, but the amount of heat is quite small because the chromosphere is so rarefied.

The outermost region of the Sun is the corona, the outer part of the Sun's atmosphere. It extends for millions of miles into space; in fact, we can think of the Earth as revolving around the Sun in the Sun's outer corona. The density of this part of the Sun's atmosphere, however, is less than the finest laboratory vacuum. The temperature of the corona is more than a million degrees. Again, there is little heat because there is so little matter in the corona. There are only a billion atoms of matter per cubic centimeter in the Sun's inner corona compared to 10 billion times more atoms in the same volume of the Earth's atmosphere at sea level.

The Sun's corona can only be seen during a total eclipse or in a coronagraph, a device that creates an artificial eclipse of the Sun inside a telescope, because the light from the corona is overwhelmed by light from the photosphere. The photosphere, chromosphere, and corona together are regarded by solar astronomers as the Sun's atmosphere.

In the photosphere there are regions, called granules, that are 300 to 1,000 kilometers in diameter, in which hot gases rise from regions below the photosphere. These gases cool, become denser, and sink below the photosphere at the edge of the granules. Imposed on these granular cells are supergranulations, regions in the photosphere that average 30,000 kilometers in diameter, in which hot gases move horizontally from the center to the edge. The entire photosphere also pulses up and down in regions that are 3,500 to 7,000 kilometers across, taking about five minutes for a cycle. This phenomenon has been likened to the irregular movement of waves in a choppy sea. Furthermore, the entire Sun appears to pulsate with a cycle that takes an hour or two to complete.

We can summarize our understanding of the Sun to this point by suggesting that the Sun is a dynamic, ever-changing mass of gas that is in constant flux and turbulence.

The Sun: Surface features

Sunspots are the most conspicuous features in the Sun's photosphere; a few of them are large enough to be seen with the unaided eye. Long before the invention of the telescope the Greeks and other ancient peoples knew about sunspots. Even though we cannot, and should not try to, look at the Sun without proper protection, the Sun can occasionally be viewed directly in the early morning or late evening because its light comes through so much atmospheric haze. It was at these times that ancient peoples saw enormous sunspots.

Sunspots are regions of gases that appear darker because they are about 15 hundred degrees Kelvin cooler than the surrounding photosphere. If we could isolate sunspots and put them in space they would be quite bright because their temperatures are about three times higher than the temperature inside a furnace where steel is made.

The cooler gases flow away from the center of the sunspot and are then heated by the hotter gases of the photosphere. Sunspots last from a few hours to a few months. Quite frequently they appear in groups of 2 to 20

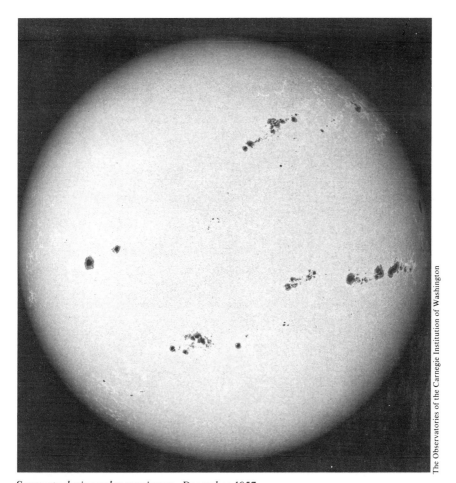

The Observatories of the Carnegie Institution of Washington

Sunspots during solar maximum, December 1957.

spots. Generally there is a dominant spot and a smaller, following spot. The largest spots may be 50 thousand kilometers in diameter, much larger than the Earth. The Sun's rotation carries the spots from west to east; indeed, Galileo first showed that the Sun rotates by watching sunspots day after day.

Every 11 years there is a sunspot maximum; halfway between is a sunspot minimum. There may be one hundred or more significant spots on the Sun during a period of sunspot maximum and no spots at all during a minimum period. In the years between 1645 and 1715 there were few spots on the Sun; the 11-year cycle apparently was nonexistent. It is perhaps no coincidence that these were the same years known as the Little Ice Age. Thus, there seems to be some cause-effect relationship between solar activity, part of which is displayed by sunspots, and climate conditions on the Earth.

There are other spectacular phenomena on the Sun: plages, bright regions emitting light from calcium and hydrogen atoms; spicules, jets of gas shooting upward at 30 kilometers per second from 5 thousand to 20 thousand kilometers above the photosphere; flares, very bright but short-lived events that put out tremendous amounts of energy; and prominences, those magnificent, graceful loops and arches of gas that sometimes extend hundreds of thousands of kilometers above the photosphere. All of these events, together with sunspot activity, are related to strong localized magnetic fields that drive the solar phenomena cycle.

It is not yet completely clear how these local magnetic fields are related to solar phenomena, but they seem to be tied in with the differential rotation of the Sun. That is, because the Sun rotates at different speeds in different latitudes, changes occur in local magnetic fields that in turn influence sunspots and these other phenomena.

Supernovae: Neutron stars and pulsars

We are not able to make direct observations of stars going through changes during their lives because they evolve over long periods of time—millions to billions of years. Occasionally, however, a star near the end of its life span will explode violently because it has collapsed due to its own gravitation. The change in brightness may be great enough to be seen with the unaided eye. We call this event a supernova, or super new star. It really is an old star that is brighter simply because it has suddenly erupted and thrown off material in all directions in space. Only seven naked-eye supernovae have been recorded in the past few thousand years.

An exploding star will typically become very bright in a few hours or days and then gradually fade during a period of several weeks or months. Left in the center of the supernova is a hot, dense core of material. If the star started with a mass several times the mass of the Sun, the core that is left becomes a neutron star. The particles in a neutron star are so squeezed together by the strong gravitational field that atoms can no longer exist. Electrons, protons, and neutrons—the particles that are the building blocks of atoms—are the dominant forms of matter. Indeed, these particles are so close together that the electrons, which have a negative electrical charge, are squeezed onto the protons, which have a positive electrical charge. The result is a neutron, a particle (as the name implies) with no electrical charge. A tablespoon of the material of a neutron star might weigh millions of tons because the particles of matter are so closely compacted.

If a neutron star is spinning rapidly, we then call it a pulsar. Pulsars, first discovered in 1967 by English astronomers, emit pulses of radio energy and thus can be detected by radio telescopes. The pulsars that have been discovered (about three hundred to date) have periods ranging from about one pulse every three seconds to one pulse every 16/10,000 of a second. This latter pulsar is thus spinning 625 times per second. The rate of pulses from some of these objects also seems to be decreasing.

It is not yet clear how neutron stars emit their pulses. There may be a "hot spot" on the surface of the star that gives off energy; every time it rotates we record this energy as a pulse or burst of radio energy. Alternatively, a neutron star may have a large magnetic field that captures electrons that have velocities close to the speed of light. As these particles move along the lines of the magnetic field they give off energy. If the star is rotating rapidly, the lines of the magnetic field might cross our line of sight and the energy given up by the particles in the magnetic field would be recorded as pulses of radio and other kinds of electromagnetic energy.

Neutron stars and pulsars are thought to be no more than a few kilometers in diameter. Thus, if we start with a star that is larger and more massive than the Sun, squeeze most of its matter into a sphere that has a diameter no larger than the distance across a medium-sized city, and spin it on its axis as fast as 625 times each second, we then have a neutron star or pulsar.

Perhaps the most famous supernova event that has a pulsar at its core is the Crab Nebula. It was a supernova explosion that was recorded in the year 1054 by Oriental astronomers and possibly by native Americans in the desert Southwest. The core pulsates once every 0.033 second in both the optical and radio wavelengths of energy.

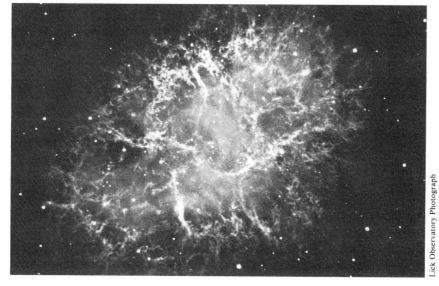

The Crab Nebula in the constellation of Taurus. This supernova remnant is still expanding at hundreds of kilometers per second after the explosion of a massive progenitor star more than nine hundred years ago.

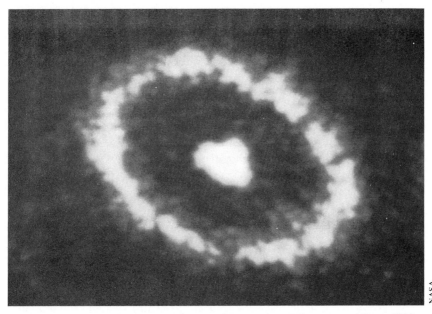

NASA

*Supernova 1987A, photographed by the Hubble Space Telescope in August 1990.
The elliptical, luminescent ring of gas is 1.3 light-years across and is the relic of the
hydrogen-rich stellar envelope ejected by the progenitor star an estimated 10
thousand years before the supernova explosion. The ring, still with a temperature
of 20 thousand degrees Kelvin 3¹/₂ years after the supernova event, fluoresces due
to ultraviolet light from the supernova explosion. The supernova explosion itself is
the asymmetrical configuration in the center of the ring. It is still too dense to
reveal whether the central object is a neutron star, pulsar, or black hole.*

The most recent supernova occurred in February 1987 at a distance of 170
thousand light-years in the Large Magellanic Cloud. A twelfth magnitude star
suddenly increased its brightness by four thousand times to about third
magnitude. It could thus be seen by the unaided eye, but only from the
Southern Hemisphere.

If a star at the end of its lifetime contains even more mass, say on the order
of eight times the mass of the Sun, it will again collapse in a supernova event,
but, because there is so much more mass involved, not even neutrons can
withstand the gravitational pressure of collapse. The core continues to
collapse until it becomes a black hole.

Telescopes

Telescopes with which we are familiar collect visible wavelengths of light
from the spectrum. There are two basic kinds of telescopes: the refractor
and the reflector.

The refractor has a primary lens, called the objective, made of clear optical glass which is usually convexly curved on both sides. Light is bent, or refracted, through the glass to bring it to a focus at the focal point. The distance between the primary lens and the focal point is called the focal length. The eyepiece consists of another refracting lens that magnifies the image brought to the focal point by the primary lens.

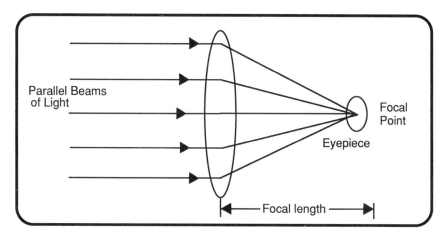

Many people inquire about how powerful a telescope is. Power is expressed by how many times the image is magnified and is obtained by dividing the focal length of the objective lens by the focal length of the eyepiece lens. Thus, if we have a primary lens with a focal length of one hundred centimeters and an eyepiece with a focal length of one centimeter, the power of the telescope is one hundred, usually expressed as 100 X, or one hundred power. If we change eyepieces—and even small, inexpensive telescopes are usually equipped with two or three eyepieces of various focal lengths—to one with a focal length of $1/2$ centimeter, we are then magnifying two hundred times ($100/1/2 = 200$), or we have a telescope of two hundred power.

The largest refractor in the world is the 40-inch telescope in the Yerkes Observatory in Williams Bay, Wisconsin. The 40-inch designation means that the primary lens is 40 inches in diameter. It is doubtful that a larger refractor will ever be built because there is no way to support the primary refracting lens except at its edge. This arrangement makes it hard to maintain accurate optical surfaces on the lens.

The other basic kind of telescope is the reflector. The primary lens is a concave piece of glass or other material that will not be greatly affected by changes in temperature, which would distort the accurate optical surface. Light is reflected from a thin layer of aluminum or some other highly reflective and nontarnishing material that is coated on the optically accurate surface of the glass. The reflected light is focused at the focal point where it can be magnified by a suitable eyepiece.

Frequently, and particularly in smaller telescopes, a secondary mirror with an optically flat surface interrupts the converging beams of light before they reach the focal point. The second mirror redirects the light outside the telescope where it can be more conveniently studied. This arrangement is called the Newtonian Telescope, after Isaac Newton who built the first example of this optical system. (Newton developed the first reflecting telescope in 1668. The first refractor used for astronomical purposes was invented by Galileo in 1609.)

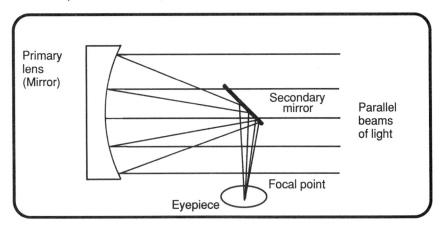

The largest reflector is the 6-meter (236-inch) telescope recently put into service in Russia. The largest reflector in the United States is the 200-inch Hale Telescope on Palomar Mountain in California. Large reflectors, at least compared to refractors, are possible because the primary lens, a mirror, can be supported from the back to maintain an accurate optical surface.

Nearly all large telescopes are used almost exclusively with some sort of electronic detecting equipment. No astronomer works visually any more, and few astronomers even use film today, because electronic sensors can capture more light than the human eye or photographic film can. Furthermore, the light can be stored as numbers in a computer and studied at leisure and in detail, a procedure not possible with the human eye as the receiver of the light.

Larger telescopes are built simply because they gather more light than smaller ones. For example, a six-inch telescope can gather more than five hundred times more light than the human eye. The two hundred-inch telescope can gather more than 11 hundred times the amount of light that a six-inch telescope can and more than 640 thousand times the light that the human eye can collect. Since astronomers work most of the time with exceedingly faint objects and small amounts of light (except for the Sun, moon, and planets), the larger the telescopes they use, the more light they can gather and the more knowledge they can derive about the objects in the Universe they are studying.

Another reason to build larger telescopes is to increase their resolving power, which is the ability of an optical system to separate two objects in the sky. The greater the resolving power, the finer the detail that can be studied. For example, a six-inch telescope can in principle resolve two stars that are 0.7 arc second apart. The two hundred-inch telescope can in principle resolve two stars that are 0.02 arc second apart. Of course, these theoretical values are never achieved because turbulence in our atmosphere limits "seeing" at even the best mountaintop sites to about one arc second.

To put these resolving power numbers into perspective, 0.7 arc second is the angle made by a quarter seen face-on at a distance of five miles. The two hundred-inch telescope could theoretically resolve the angle made by a quarter at a distance of 163 miles.

People also ask how far a telescope can see. Obviously, the answer depends on the brightness of the object being studied. A faint object relatively close to the Earth may escape detection in all but the largest telescopes. On the other hand, a very bright object, billions of light-years distant, may be available in even modest-sized telescopes. For example, some of the brightest and most distant objects in the Universe are quasi-stellar radio sources (quasars). If one knows where to look, the brightest of them can be seen in telescopes of modest aperture.

The speed of a telescope's optical system is also important to astronomers. It is expressed as the *f*-ratio of the telescope, the same designation that most of us are familiar with in our cameras. The speed of a lens is simply the focal length of the lens divided by the diameter of that lens. For example, if we have a lens that has a diameter of 25 centimeters and a focal length of 100 centimeters, the speed of the lens is *f*4. One can change the *f*-ratio on a camera not by changing the focal length of the lens but by changing the diameter of the lens using an iris or a diaphragm.

The sharper the curvature of the lens, either a mirror or a refracting lens, the shorter the focal length and the "faster" the telescope. This simply means that a lens with a sharp curvature gathers light from more of the sky; thus, if we were to take a picture, we would not have to expose the film very long. That is, we have a fast telescope. Small *f*-ratio numbers indicate a fast optical system due to the short focal length of the lens.

Slower telescopes have less curvature to their lenses, thus covering a smaller part of the sky. In other words, less light is gathered because the focal length is long compared to the diameter of the objective.

A telescope with an *f*-ratio of 15, for example, is a rather slow system. An *f*2 lens system is very fast and is used in Schmidt telescopes for survey work because so much of the sky is covered at one time. Slow systems are used to study the details of bright objects such as the Sun, moon, and planets. Many amateur telescopes work between $f8$ and $f11$, a compromise between a fast optical system that covers a lot of the sky and has images that are small and lack detail and a slow optical system that reveals larger and more detailed images.

A new development in optical astronomy allows the equivalent of very large telescope mirrors to be built. These are segmented mirrors, many large mirrors that together make one huge mirror. Light from each of the large mirrors is brought to a common focal point because each mirror has its optical surface controlled by a computer. The Keck Telescope on Mauna Kea in Hawaii, with an effective mirror diameter of 10 meters, operates with this kind of active optics.

Telescopes in space

The last decade of this century, indeed, the end of the millennium, can be called the golden age of astronomical exploration from space. Following is a selected list of scientific satellites that have been, or will be, launched into Earth orbit or sent to the planets.

Spacecraft	Mission	Year of Launch
Magellan	Map Venus	1989
Cosmic Background Explorer (COBE)	Analyze cosmic background radiation from Big Bang	1989
Galileo	Analyze Jupiter and several of its moons	1989
Hubble Space Telescope	Study planets, stars, galaxies, gas and dust clouds. Most sophisticated scientific instrument ever built	1990
Roentgen Satellite (ROSAT)	Study stellar and galactic x-rays	1990
Astro-1	Study stellar and galactic ultraviolet and x-ray sources	1990
Gamma Ray Observatory	Find and study sources of gamma rays	1990

Ulysses	Investigate the Sun and its environment from solar polar orbit	1990
Atmospheric Lab for Applications and Science	Study interaction of Sun and Earth's atmosphere	1991
Extreme Ultraviolet Explorer	Make a catalogue of stellar and galactic ultraviolet sources	1991
Mars Observer	Study Mars for possible manned flight	1992
High-Energy Astrophysics lab	Study cosmic x-rays	1992
Gravity Probe	Test of Einstein's General Theory of Relativity	1993
Advanced X-ray Astronomy Facility	Study cosmic x-rays	1995
Comet Rendezvous Asteroid Flyby	Comet and asteroid probe	1996
Earth Observing System (EOS)	Study all phases of Earth's environment from Earth orbit	1997
Cassini	Study Saturn	1998
Space Infrared Telescope Facility	Investigate Milky Way Galaxy in the infrared	1999

In addition, the decade of the 1980s saw the highly successful Voyagers 1 and 2 travel to Jupiter, Saturn, Uranus, and Neptune. The IRAS (Infrared Astronomical Satellite) found more than two hundred thousand sources of infrared radiation. The Giotto spacecraft travelled to within a few hundred kilometers of the core of Halley's comet. The International Ultraviolet Explorer (IUE) and the High-energy Astronomical Observatory (HEAO) 2, called Einstein, explored the x-ray and ultraviolet parts of the spectrum not available on the surface of the Earth. All parts of the spectrum have been, or will be, explored and will continue to be analyzed in greater detail.

We have learned more about the Universe in the last few years than we learned in the entire previous history of science. This is indeed the most exciting time to be an astronomer or to have curiosity about the astronomical Universe.

Tides

Tides are the result of the Sun's and moon's gravitation acting on the Earth's atmosphere, oceans, and land masses. We will be concerned only with ocean tides in this discussion.

The moon, much less massive than the Sun but also much closer to the Earth, exerts a greater pull on the Earth's oceans than does the Sun. Ocean water under the moon is attracted toward the moon. The mass of the Earth's solid material under the moon is also attracted, but the water is more fluid and thus flows over the Earth's surface toward those areas below the moon. This flow of water creates a bulge of water under the moon that we call a high tide.

It turns out that there is also a tidal bulge on the opposite side of Earth from the moon. This bulge is caused by the moon's gravitation pulling the solid material of the Earth on the other side toward the moon and away from the water. Thus, water flows away from the moon and bulges on the side of the Earth opposite the moon. As the Earth rotates, land masses move into and away from the tidal bulges, creating high and low tides. Differences in heights of the same high tide are the result of differences of shoreline formations.

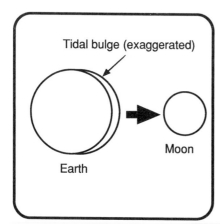 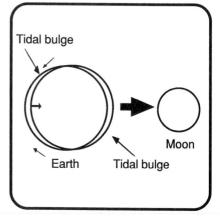

The Sun also creates tides, but only about half as large as the tides caused by the moon. When the Sun and moon are on the same side of the Earth (the moon is then said to be in its new phase) or when the Sun and moon are on opposite sides of the Earth (full moon), tides are called spring tides and are higher than at any other time. (Spring tides have nothing to do with the season of the year of the same name.) The reason is that the Sun's and moon's gravitational fields reinforce each other. When the moon is at first or last quarter, we experience neap tides, those high tides that are lower than normal because the gravitational fields of the Sun and moon work against each other.

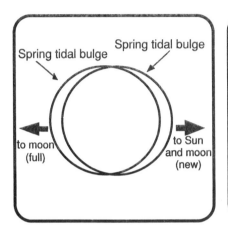
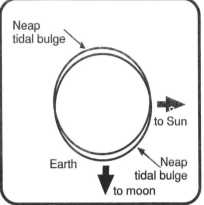

The friction between the water and the solid Earth, particularly in the shallow coastal areas, has caused the Earth to slow its rotation. It is estimated that the current rate of slowing is about 2/1,000 of a second per century. Thus, days are gradually becoming longer. If the Earth's rotation is slowing and in order to meet the requirements of the law of conservation of energy, the moon has to move away from the Earth. The present rate at which the moon spirals away from the Earth is about three centimeters per year.

The Earth exerts a much greater tidal attraction on the moon than the moon exerts on the Earth. During the several billion years that the Earth-moon system has probably existed, the Earth's gravitational pull on the moon and the consequent tidal bulge in the solid material of our satellite has slowed its rotation so that the same side of the moon now faces the Earth at all times.

Timekeeping

As the Earth rotates on its axis, objects in the sky cross the meridian, an imaginary line that runs directly north and south through the sky for every observer. When the Sun is squarely on the meridian, it is noon for the observer. Obviously, it takes 24 hours for the Sun, always appearing to move to the west through the sky as the Earth rotates on its axis, to return to the meridian. That is, it takes the Earth just 24 hours to rotate once on its axis.

Instead of reckoning the beginning of a new day when the Sun is on the upper part of the meridian, a practice that would make our daily lives difficult, by common agreement throughout the world a new day begins when the Sun is on the lower part of the observer's meridian. The time for the observer is midnight. To separate morning from afternoon—that is, before and after the Sun crosses the observer's meridian—we designate morning as A.M., which stands for *ante meridiem*, a Latin phrase for before midday. After noon we designate as P.M., *post meridiem*, after midday.

The Sun we see in the sky is called the apparent Sun and keeps apparent solar time. The Sun we see, however, does not move through the sky at a uniform rate. Part of the reason is due to the oval shape of the Earth's orbit; when the Earth is closest to the Sun, in January, it moves faster around the Sun, making the Sun appear to move faster in the sky. When the Earth is farthest from the Sun, in June, the Sun appears to move more slowly in the sky. There is only one timepiece that accurately keeps apparent solar time, and that is the sundial.

Obviously, if we want to keep accurate regular time, we need a different kind of timepiece. Astronomers have invented a fictitious Sun, called the mean Sun, which runs steadily with no change of pace such as the apparent (real) Sun displays. Our watches are set to the mean Sun, and thus tell us what the local mean solar time is.

We mentioned that every observer has his or her own meridian. Thus, the Sun can only be on one meridian at a time. Put another way, it can be noon at only one location on the Earth at a time, or, every location on the Earth has its own local mean solar time. Because it would be awkward to have a different time in every location, by mutual agreement people have adopted 24 time zones around the planet. Every place within the same time zone has the same time, although the Sun is in a different location relative to every observer's meridian. This last point is easily illustrated. If you live on the eastern edge of a time zone, the Sun is up earlier in the morning than it is for those people living on the western edge of the same time zone. Conversely, the Sun sets in the west at a later time for people living on the western edge of a time zone than it does for people living on the eastern edge of the same time zone.

Illinois, for example, is in the central time zone. Locations east of Illinois are in the eastern time zone and therefore an hour later; people in the mountain time zone are an hour behind Illinois. If we move west, we move into earlier time zones. If we go far enough to the west, we could move into yesterday! To avoid this impossible situation of moving into the past (or future if we travel east), the international date line was established. When one is moving west and crosses this imaginary line, the date changes from today's date to tomorrow's date. If we travel from west to east across the international date line, we set our calendars back one day to yesterday.

Finally, we have installed daylight saving time in the summer, an invention that still confuses people. We are not really saving time; all we are doing is moving our clocks ahead by one hour in spring so that the Sun appears to rise and set one hour later. We have traded an hour of daylight in the early morning, when most of us are sleeping anyway, for an hour of daylight in the evening so that people have enough daylight at the end of their working day to cut grass, tend gardens, and wash cars. In the fall, we set the clocks back one hour, trading an hour of daylight in the evening for an hour in the morning. We are then again on standard time.

Unidentified flying objects

The Universe is mostly space that is nearly empty; distances between stars are enormous. Light, which moves faster than anything else in the Universe and whose speed cannot be exceeded, takes more than four years, traveling at 300 thousand kilometers (186 thousand miles) per second, to reach the Solar System from the next closest star. Light from the next closest spiral galaxy to the Milky Way requires more than two million years to reach us.

These facts—the near emptiness of space, the enormous distances between stars and galaxies, and light as the limiting speed of the Universe—all provide strong arguments against the idea that the Earth is being visited by intelligent creatures from planets orbiting stars beyond the Sun. (We are reasonably certain that intelligent life exists in our Solar System only on the planet Earth. The next best candidate for life of any kind in the Solar System is Mars. The two Viking spacecraft on the surface of Mars were designed to detect life; they found nothing.)

It seems highly unlikely that intelligent life elsewhere in our galaxy or beyond would single out our star, the Sun, from the four hundred billion or so stars in the Milky Way to search for life on one of the Sun's planets. The problem is one of finding planets in the first place. Even with our sophisticated and sensitive telescopes and other instruments, we have no direct evidence of planets orbiting stars beyond our own Solar System; the stars are simply too far away. The chances of the Earth being found are equally remote.

No mass, such as a spaceship, can travel at the speed of light; only light can do that. Thus, if life on our planet were detected by intelligent life on a planet beyond the Solar System, or if we could detect intelligent life elsewhere, distances and therefore travel times would be prohibitive.

Furthermore, space between stars—and probably between galaxies, too— is not quite empty. The density of matter is much less than our finest laboratory vacuums. Nevertheless, the few atoms and molecules in interstellar and intergalactic space present a barrier for mass traveling close to the speed of light. A spacecraft traveling close to the speed of light would hit enough of these molecules that it would burn up due to the friction that would be created. Thus, speeds needed to travel between stars or galaxies seem to be impossible to maintain. Put another way, given the laws and conditions of the Universe as we understand them, manned exploration of space seems pretty well confined to our own neighborhood; we will never travel to stars beyond our Solar System.

There are a great number of celestial phenomena that astronomers and other scientists have not been able to explain yet. We must look, however, for the most probable—not the least probable—explanations. The least probable explanation of all UFO observations is that the planet Earth is being visited by intelligent life in spaceships. As a matter of fact, this explanation is so improbable that we may discount it altogether.

Science and astronomy cannot prove that something, such as spaceships from other planets, does not exist. All science, including astronomy, deals only with the observable and measurable. Until a spaceship from beyond the Earth is available for scientific scrutiny and analysis, we must turn to more likely explanations for the many phenomena that are observed in the sky.

Finally, we should note that life, including intelligent life (however we may define it), may be a common condition throughout the Universe. Simply stated, conditions in the Universe impose limits on making physical contact between intelligent life forms. But whether we are alone in the Universe or whether there is life elsewhere, either circumstance is both frightening and awe-inspiring.

Uranus

William Herschel discovered the seventh planet from the Sun in 1781 through a six-inch telescope. The planet is named for Urania, the muse of astronomy in classical mythology.

Uranus has a mean distance of 2,875,000,000 kilometers from the Sun; at this distance it takes 84 Earth years to revolve once around the Sun. The plane of Uranus's orbit very nearly coincides with the plane of the Earth's orbit.

Much larger than the Earth with a diameter of 51,118 kilometers (compared to 12,756 kilometers for the Earth), Uranus has 14.5 times as much mass as the Earth. It rotates in about 17 hours. Because its axis is tipped so that it is almost in the plane of its orbit, Uranus can be thought of as rotating in a retrograde, or opposite, direction from the direction of rotation of the Earth. Uranus's density is slightly more than the density of water. The planet has a strong magnetic field that is tipped 60 degrees to its axis of rotation, a highly unusual circumstance compared to the other planets. The magnetic field is also offset from the core of the planet.

Uranus is easily seen through a pair of binoculars or a small telescope. It appears greenish due to the small amount of methane in the planet's atmosphere. Hydrogen is the most abundant element in the planet, with helium the second most abundant.

Voyager 2 orbited close to Uranus and revealed few distinctive features in the planet's atmosphere. The minimum temperature at the top of the clouds in Uranus's atmosphere is about 52 degrees Kelvin (366 degrees below zero Fahrenheit). Uranus has a high albedo, or ability to reflect light. About 66 percent of the visible rays of sunlight that strike the planet are reflected into space.

Fifteen moons revolve around Uranus; none of them is as large as the Earth's moon. One of the moons, Miranda, a mass of frozen water and rock, presents one of the most complex topographies of any object in the Solar System.

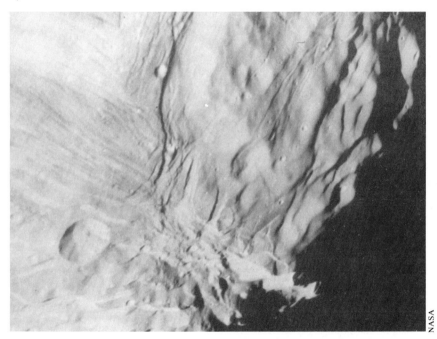

High-resolution photograph of Miranda, Uranus's most intriguing satellite. Craters, ice mountains, and other complex surface features are apparent in this image from Voyager 2.

In addition, Uranus has 11 rings that are quite dark and consequently reflect little light. The rings are made of a large number of small chunks of frozen gases, each in its separate orbit around the planet. In this way they resemble the rings of Saturn.

Venus

After the Sun and the moon, Venus is the brightest object in the sky. Venus is often called the sister planet to the Earth because it is almost the same size as the Earth; it has a diameter of 12,104 kilometers, compared to 12,756 kilometers for the Earth. It also has about 80 percent of the mass of our planet. Its density, too, is more than five times that of water, comparable to the density of the Earth. Its surface gravity is 88 percent that of the Earth's.

There are significant differences between Venus and the Earth, however. Venus is only 109 million kilometers from the Sun, taking just 225 Earth days to revolve once around our star. Venus rotates once on its axis in 243 Earth days and has a retrograde motion; that is, it rotates in the opposite direction from the rotation of the Earth and all the other planets except Uranus. Thus, if the Sun were visible through the Venusian clouds, it would rise in the west and set in the east.

Even though Venus comes closer to the Earth than any other planet, its surface is not visible from the Earth because the planet has an atmosphere heavily laden with clouds. The atmosphere is composed almost entirely of carbon dioxide, with trace amounts of carbon monoxide, water vapor, and oxygen. Due to sulfuric acid in the clouds, they are slightly yellow in color. Hydrogen chloride and hydrogen fluoride may also be present. When these compounds combine with water they form highly corrosive acids.

The atmosphere is so thick that at Venus's surface the pressure is 90 times the atmospheric pressure at the Earth's surface. The temperature at the top of Venus's clouds is about 235 degrees Kelvin (36 degrees below zero Fahrenheit), but at the surface of the planet the mean temperature is 730 degrees Kelvin (855 degrees Fahrenheit). This very high temperature—hot enough to melt soft metals such as tin and lead—is the result of the greenhouse effect of Venus's atmosphere. Some light from the Sun is able to penetrate the atmosphere. When the light tries to get back out after being reradiated from the surface of the planet as heat, however, it is absorbed by the sulfuric acid and carbon dioxide. Thus, the clouds are more effective in blocking the energy escaping from the planet than they are in preventing it from reaching the surface, and the surface is consequently quite hot. If we were to stand on Venus, the brightness from the sunshine would be comparable to a dark, rainy day on the Earth.

Even though no one has seen the surface of Venus directly from the Earth due to the thick clouds, Russian space probes have landed and sent back pictures of the surface. They show rocks up to 70 centimeters (more than two feet) in diameter, some with sharp edges. They appear to rest on a dark material. In addition to these images from the surface of Venus, radar mapping of the planet indicates that the surface consists of high plateaus,

NASA

This mosaic of 56 pictures of Venus shows the thick carbon dioxide atmosphere and clouds that block direct inspection of the surface of Venus in visible light.

volcanoes, craters, and valleys—not unlike many of the surface features on the Earth or moon.

The interior of the planet has been assumed to be similar to that of the Earth. That is, it is probably a liquid, metallic core surrounded by a mantle and a rocky crust.

White dwarfs

When the Sun has converted most of its hydrogen to helium in the nuclear fusion furnace in its core, it will contract, then expand and thus gently explode, throwing matter into all directions in space. The core that is left will be extremely dense—up to a million times denser than water. One tablespoon of this matter, if we could find some method of putting it on a scale, would weigh one thousand tons. Cores of stars that remain after they explode at the ends of their lifetimes are called white dwarfs because they are small (compared to their original sizes) and because they are white hot.

The density of matter in a white dwarf is far greater than the density of any solids with which we are familiar in our everyday experience on Earth. Matter can be squeezed together in white dwarfs because atoms are mostly empty space. Temperatures and pressures are very high in white dwarfs; thus, most of the electrons have been stripped from their atoms. These free electrons form a degenerate gas that cannot be squeezed further by the star.

White dwarfs are the end result of stars that start out with masses between about $1/5$ and 1.4 solar masses. Many, if not a majority, of the stars we see in the nighttime sky fall within these limits. This is to say that many stars end their lives as white dwarfs.

If a star starts with a solar mass only $1/5$ that of the Sun, its final white dwarf stage will have a diameter only $1/50$ that of the Sun, or about 28 thousand kilometers. If a star has 1.4 solar masses, its final white dwarf size will be only a few kilometers in diameter. The reason is that the larger the mass a star starts with, the stronger its gravitational field, and the more it will contract. If a star has more than 1.4 solar masses, it will contract beyond the white dwarf stage to a pulsar, or neutron star. If the star starts with many more times the mass of the Sun, it will contract to a black hole and disappear from the Universe.

The temperature at the surface of a white dwarf may reach 50 thousand degrees Kelvin (100 thousand degrees Fahrenheit). White dwarfs cool down, however, during a period of hundreds of millions of years and finally become dead, dark masses. Some astronomers refer to this as the black dwarf stage.

Even though white dwarfs are very hot and consequently very bright for a while, they are so small that their total luminosity is quite low. They are therefore very faint stars as we see them in the nighttime sky. Only the closest white dwarfs can be studied and there are only several hundred that have been identified.

Because so much matter is compressed into a relatively small volume, the gravitational fields of white dwarfs are very strong. They have thus been used to test Einstein's General Theory of Relativity. Light leaving any massive object has to expend energy to escape the object's strong gravitational field and therefore leaves with less energy. That is, light leaving a white dwarf has less energy and is therefore redder than it would be leaving a normal star. Indeed, these Einstein gravitational redshifts, as they are called, have been observed in the light from several white dwarf stars. The results confirm the General Theory of Relativity within the limits of observational error.

Why is the nighttime sky dark?

This seemingly simple question has a seemingly simple answer: because the Earth rotates on its axis and the Sun is shining on the other half of the planet.

But things are not so simple. Four hundred years ago Johannes Kepler asked the same question and came up with an interesting answer. He reasoned that if the Universe were infinite, it would have an infinite number of stars in it. No matter where we looked we would encounter the light of a star. Therefore, the nighttime sky should be as bright as the daytime sky. Kepler reasoned that since the nighttime sky is dark, the Universe cannot be infinite.

Wilhelm Olbers, in 1826, posed the same question. He used the inverse square law to note that the light received from stars decreases as the square of their distance from us. However, if there is a uniform distribution of stars, the number of stars increases as the square of their distance. Thus, these two circumstances cancel each other. Therefore, the nighttime sky should be bright. Stated this way, the problem is known as Olbers's paradox.

One answer that was given is that dust clouds block the light from stars behind them. Yet, this is not a good answer because the dust would eventually be heated by starlight and glow with the same temperature as the stars. Since we do not find hot dust clouds, this answer is obviously not correct.

Another proposal may be at least part of the answer. We know that the Universe is expanding. Therefore, light that would be in the visible part of the spectrum is shifted so far toward longer wavelengths as a result of the Doppler effect that most of its energy is lost. In other words, this radiation is so feeble that it is in the very long wavelengths of the spectrum, far beyond visible light.

Another answer has recently been suggested that is quite simple: there are not enough stars—far less than an infinity of stars. Much of the sky does not contain a star, a source of light. If we add up the light from all the stars we think are in the Universe, there are not enough stars to make the nighttime sky bright.

There are objections to these solutions, but perhaps astronomers are on the right track to answer that deceptively simple question: Why is the nighttime sky dark? When we know the answer to this question, we will know a great deal about the Universe.

The year and the calendar

The day, the month, and the year are all based on the motions of astronomical objects. The day is defined as that period of time that it takes the Earth to make one complete rotation on its axis with respect to the Sun.

There are two kinds of months: the sidereal, or star month, and the synodic month. The sidereal month is about $27\frac{1}{3}$ days in length and is the period of revolution of the moon about the Earth with respect to the stars. However, the moon revolves around the Earth, and the Earth revolves around the Sun. This means that a little longer time is required for the moon to move from one phase back to the same phase. For example, when the moon, Sun, and Earth are all lined up, the moon is said to be in its new phase. It takes about $29\frac{1}{2}$ days, a synodic month, for all three objects to be aligned again.

There are two kinds of years that are important. The first is the sidereal year; it is 365.256 solar days in length. It is the period of time it takes the Earth to make one revolution around the Sun with respect to the stars. The tropical year is the length of time it takes the Earth to revolve around the Sun with respect to the start of the seasons. It is 365.242199 solar days in length, or 20 minutes and 24 seconds shorter than the sidereal year.

Thus, a day is the result of the rotation of the Earth, a year is the result of the revolution of the Earth around the Sun, and a month is the result of the revolution of the moon around the Earth. (The words moon and month were originally the same word.) The reason that an accurate calendar is so difficult to devise is that the length of a year, the length of a day, and the length of a month are represented by numbers that are not evenly divisible into each other. Most modern calendars have stopped trying to reconcile the motion of the moon with the year and the day. Thus, the moon is no longer important in calendar-making except for determining a few religious holidays such as Easter and Passover.

Our present calendar goes back to the Romans, specifically to Julius Caesar who reformed the earlier Roman lunar calendar to a solar calendar in 46 B.C. The Julian calendar, as it is called, has 12 months and a total of 365 days. Every four years an extra day is added to make up for the approximate one-quarter day in every tropical year. Thus, we have 366 days in every fourth year, which we call leap year; the extra day is added to February.

By the year 1582, the Julian calendar had fallen out of step with the Sun by 10 days because there is slightly less than one-quarter day in every tropical year. Pope Gregory XIII installed another calendar reform by dropping 10 days out of 1582; October 4 became October 15 by papal proclamation. Further, in order to make the calendar and the Sun reasonably coincident, he said that century years evenly divisible by four hundred should be leap years. Thus, the year 1600 was a leap year. The century years of 1700, 1800, and 1900 were not evenly divisible by four hundred and thus were not leap years as they would have been in the Julian calendar. The year 2000 will be a leap year.

Not all countries immediately adopted the Gregorian calendar. Catholic countries took it as their calendar, but it was not until 1752 that England and America (basically Protestant countries) adopted the Gregorian calendar. The Soviet Union changed from the Julian to the Gregorian calendar after the Bolshevik revolution. In so doing the country had to drop 13 days.

The Gregorian calendar, now in use throughout the world, was slightly revised recently. The years 4000, 8000, and 12,000—all of which would have been leap years—are now simply common years. The Gregorian calendar is thus currently accurate to about one day in 20 thousand years.

The zodiac

The word zodiac means circle of animals. It is a belt of constellations in the sky through which the Sun moves during the course of a year. All the constellations represent either animals or people except Libra, the Scales. The constellations are, in order from west to east: Aries, the Ram; Taurus, the Bull; Gemini, the Twins; Cancer, the Crab; Leo, the Lion; Virgo, the Virgin; Libra, the Scales; Scorpio, the Scorpion; Sagittarius, the Archer; Capricorn, the Sea Goat; Aquarius, the Water Bearer; and Pisces, the Fish.

As the Earth revolves around the Sun counterclockwise as seen from the

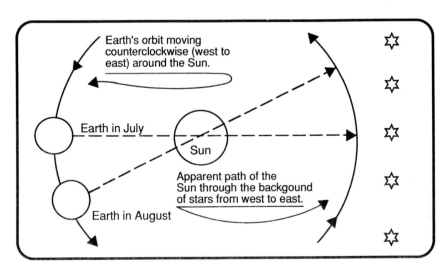

north, our perception is that the Sun is slowly moving through the background of stars from west to east. (Of course, we are not able to see the stars when the Sun is up, but we are able to calculate precisely where it is against the background of stars that are outshone by the brilliance of the Sun.)

To us on the Earth the Sun appears to move eastward through the stars that make up the constellations of the zodiac. In one year, the time it takes the Earth to make one complete revolution around our star, the Sun will appear to be back in the same place with reference to the stars. The Sun never moves out of the zodiac but simply repeats its journey through the same constellations year after year.

The moon and all the planets are also always found in the zodiac. The reason is that the orbits of the planets are centered on the Sun, of course, and none of the planes of the orbits of the planets is tilted more than 17 degrees from the plane of the Earth's orbit. Thus, the Sun, the moon, and all the planets are always located in one of the 12 constellations that make up the zodiac. None of these celestial objects will ever be found in any other part of the sky.

Zodiacal light and the gegenschein

Space between the planets in the Solar System is not empty. Tiny solid particles (as small as a 1/10,000 of a centimeter) pervade the Solar System. This interplanetary dust is found along the zodiac, which is defined as that part of the sky occupied by the planets, Sun, and moon. These dust particles are made of silicates, iron, and nickel. The mass of all the particles is thought to be on the order of 10 trillion tons.

The source of this dust is probably comets that are disintegrating in the inner Solar System, leaving their dust behind. When these dust grains hit the surface of the Earth, we refer to them as micrometeorites. The Earth gains mass due to micrometeorites that amounts to several thousand tons per day.

This dust also produces the zodiacal light, a faint cloud of light that can be seen in a clear, dark sky before the twilight that signals the rising Sun in the east and after evening twilight in the west. The sky has to be completely dark (except for starlight)—no moon, no urban or rural lights—a condition that is increasingly difficult to find anywhere on the planet. Look for a faint, triangular glow of light in the west just after evening twilight and in the east just before morning twilight. What you will see is sunlight reflecting off these interplanetary dust grains.

The gegenschein (a German word meaning opposite glow) is sunlight reflected off these same dust grains. It is seen as a very faint glow on the ecliptic opposite the Sun in the nighttime sky around midnight. The gegenschein is so faint that almost no one sees it anymore because of artificial lighting everywhere on the Earth.

We have now finished the Zs, and thus have come to the end of this book titled *Astronomy from A to Z.* If you had as much fun reading it as I have had writing it, we have all had a wonderful experience.

I would like to conclude with the motto we have adopted at the Sangamon State University Observatories:

Per Aspera Ad Astra
(Through hard work to the stars)

Constellations

The drawings that follow represent a few of the better known constellations that can be seen during various times of the year.

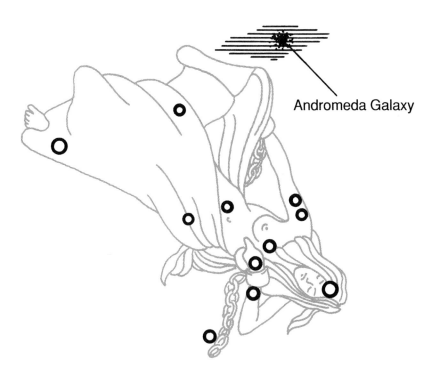

Andromeda Galaxy

Andromeda

Andromeda is high in the northeast sky about 9:00 P.M. during the fall months. In Greek legend, she is the daughter of Cassiopeia. The Andromeda Galaxy is in this constellation; it is the next closest spiral galaxy to the Milky Way, more than two million light-years away. The Andromeda Galaxy can just barely be seen with the unaided eye on a clear, dark night as a faint, fuzzy patch of light.

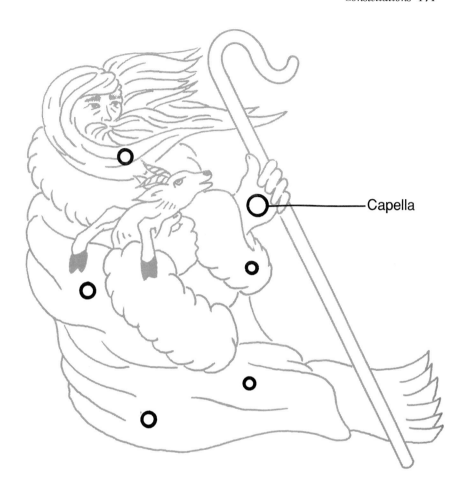

Capella

Auriga (ô-rī′gė)

About 9:00 P.M. nearly overhead from the mid-latitudes in the winter is Auriga, the Charioteer. The brightest star in the group is Capella, a yellow giant star 45 light-years from the Sun. The word Capella means she-goat; her kids are the three fainter stars of about the same brightness nearby.

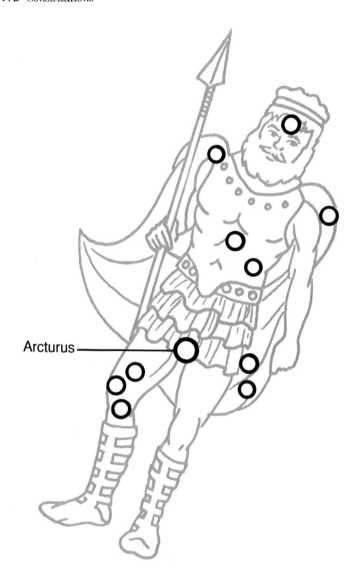

Arcturus

Boötes (bō-ō'tēz)

High in the summer western sky is the Bear Herder, Boötes. The brightest star in the constellation is Arcturus, the third brightest star in the nighttime sky. Arcturus is famous for providing the light, through a telescope, that threw the switch that turned on all the lights at the Chicago World's Fair in 1933.

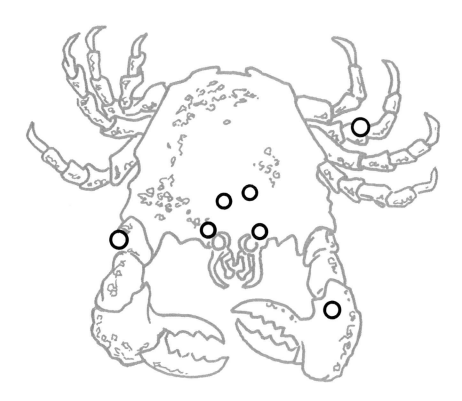

Cancer

A little more than halfway up the southern sky in early spring, between the constellations of Leo and Gemini, is Cancer, the Crab. This is a rather small and faint group of stars that the ancients saw as a crab in the sky.

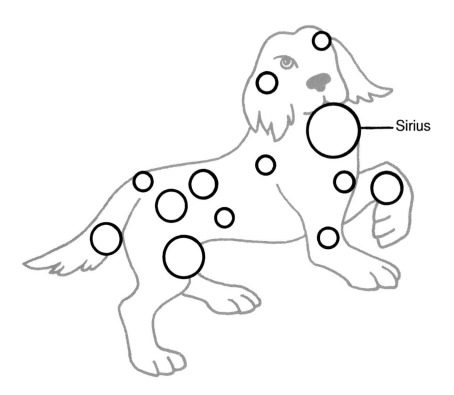

Sirius

Canis Major

Except for the Sun, Sirius, in the constellation Canis Major (the Big Dog), is the closest star visible to the unaided eye that we can see from the continental United States. Its distance is about nine light-years (approximately 85,000,000,000,000 kilometers). Sirius is also the brightest star that can be seen from anywhere on Earth in the nighttime sky. Sirius was known to the ancient Egyptians as Sothis. It appears just before the Sun rises in the east at about the time of year the annual floods occurred in the Nile valley. The Egyptians believed that Sothis caused the floods and hence was the creator of all green, growing things. To the Romans, the hottest part of the year occurred when Sirius rose close to the Sun. We still use the expression "dog days" for the hot days of August.

Cassiopeia

High in the northeast sky from dusk in the fall is the familiar "W," Cassiopeia. In Greek legend she was the queen of Ethiopia and the mother of Andromeda.

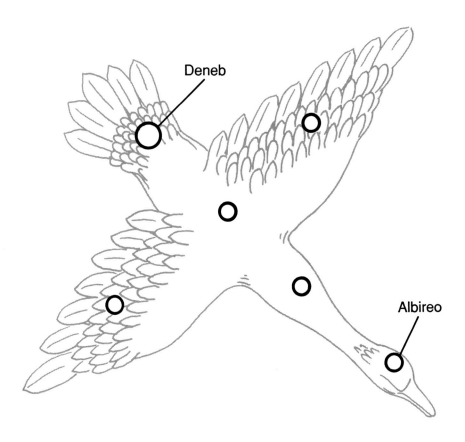

Cygnus

Cygnus, the Swan, is in the eastern sky just after sunset in early summer. It is one of the easiest of summer constellations to find because it is located in the summer triangle. The tail of the swan is the star Deneb, an Arabic word meaning tail of the hen. Its head is the star Albireo, a beautiful double star visible through even a small telescope; one star is blue, and the other star is yellow. Cygnus is sometimes called the Northern Cross, with Albireo as the foot of the cross and Deneb as the top of the cross. The swan's wings form the crossbar.

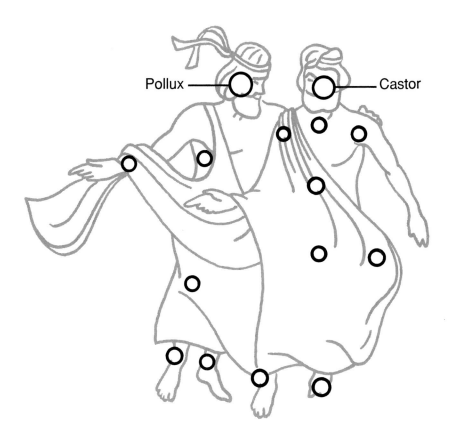

Gemini

Look southwest about 9:00 P.M. in the early spring. You will see two bright stars that shine with an equal amount of light. These two stars are Castor and Pollux, the heavenly twins, in the constellation Gemini. Castor is a multiple star system, as seen through a telescope, of at least six stars. The system is about 49 light-years (4.6 X 10^{14} kilometers) from our Solar System.

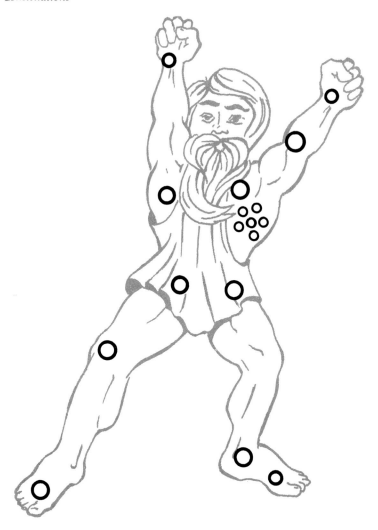

Hercules

This rather faint but large constellation is high overhead just after dark in midsummer. It represents the strong man of ancient mythology who accomplished the famous 12 labors of Hercules. Located in this constellation is the Hercules star cluster, barely visible to the unaided eye on a clear night but a magnificent sight even through a small telescope.

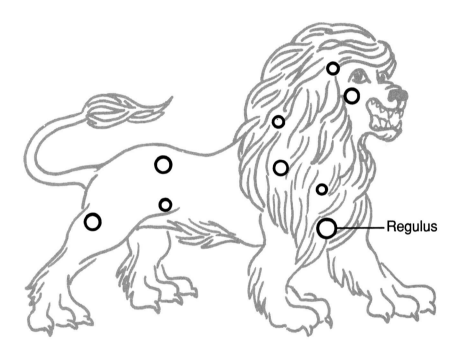

Regulus

Leo

Just above the eastern horizon about the middle of the evening in midwinter is Leo, the Lion. In Greek mythology, Leo was killed by Hercules and raised to the sky by Jupiter in honor of the hero. The constellation is easily recognized by the backward question mark of stars that outline the lion's mane. At the base of the question mark is the bright star Regulus, 69 light-years from the Sun.

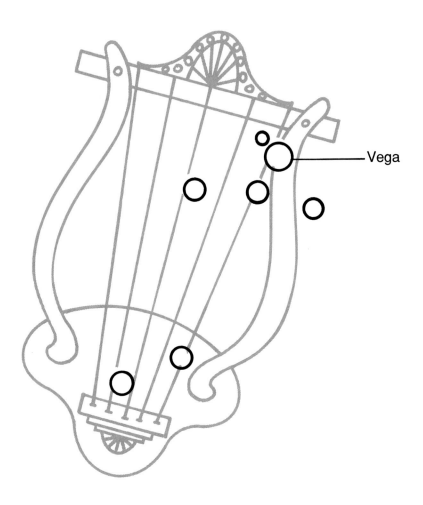

Vega

Lyra

The name of this constellation means harp. It is a small group of faint stars visible high overhead in midsummer just after dark. The constellation is easy to find, however, because it contains one of the brightest stars in the sky, Vega. The Ring Nebula is also in Lyra. It is an exploding star that is one of the most beautiful sights in the nighttime sky, even through a small telescope.

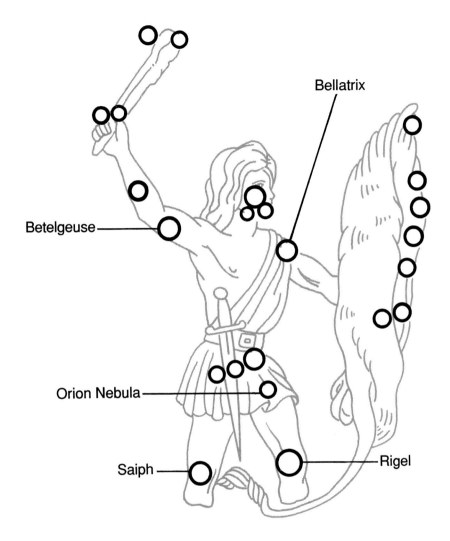

Orion

High in the southeastern sky at sunset in midwinter is Orion. He is the mighty hunter of the sky in ancient mythology, challenging Taurus, the Bull, with club and shield. The star Betelgeuse in his right shoulder is so large that if we put the Sun in its center, the edge of the star would engulf the orbit of the planet Earth. Rigel, in Orion's left knee, is a white-hot star, having a surface temperature approaching 25 thousand degrees Kelvin.

Hanging below the three stars in Orion's belt is the Orion Nebula, a large mass of gas in which new stars are being formed. This object is a beautiful sight through a pair of binoculars or a small telescope.

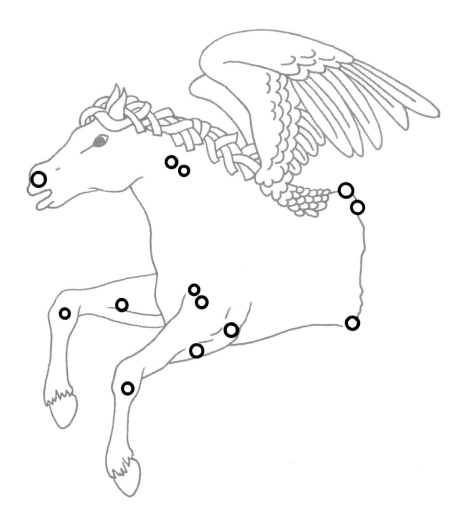

Pegasus

High in the eastern sky in fall is Pegasus, the winged horse of mythology. This constellation is easily recognizable by the great square of four stars, all about the same brightness and distance apart as seen from the Earth, that makes up the front part of the horse.

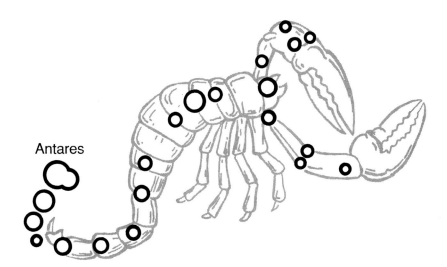

Antares

Scorpius

Just over the southern horizon after dark in the summer is Scorpius, the Scorpion. You will see a bright red star, Antares, the stinger in the scorpion's tail. Antares is a red giant star, several hundred times the size of the Sun and 25 thousand times brighter. Just to the east of Scorpius is the center (nucleus) of the Milky Way Galaxy.

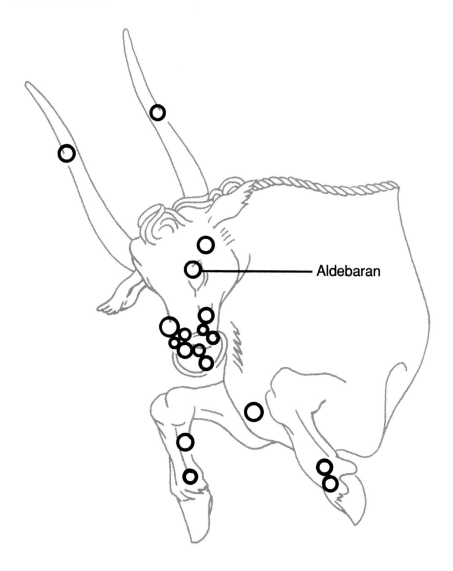

Aldebaran

Taurus

Rising over the eastern horizon in the middle of the evening in late fall is Taurus, the Bull. This constellation is well known for the bright red star Aldebaran (Ăl-dĕb′a-răn). This star, the eye of the bull, is in a V-shaped group of stars, called the Hyades (hī′e-dēz′), that outlines the bull's face. Aldebaran is 50 times the size of the Sun and is 60 light-years away from the Solar System.

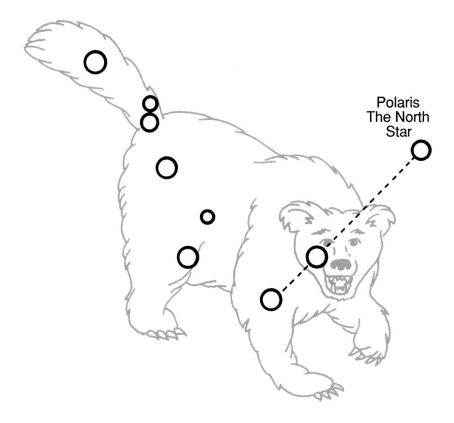

Polaris
The North
Star

Ursa Major

The stars that form the Big Bear, also known as the Big Dipper, are perhaps the most famous group of stars in the nighttime sky. Using the pointer stars, we can easily locate Polaris, the North Star. Polaris never moves and thus is a constant guide to north for us.

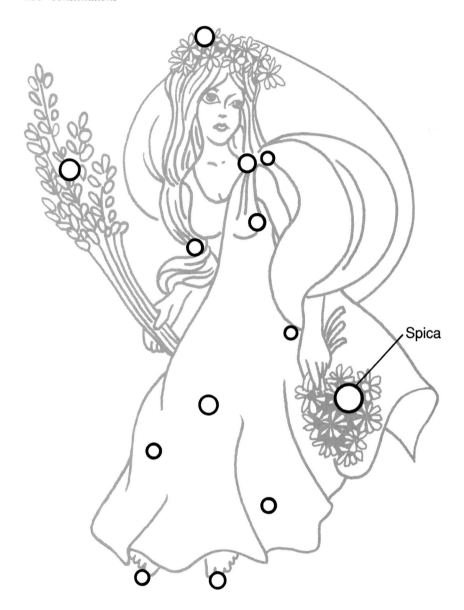

Spica

Virgo

This constellation is found in the southeastern sky at 9:00 P.M. in the spring.
It represents a young woman carrying a sheaf of wheat or stalk of corn. The
constellation is recognizable because it contains one bright star, Spica,
meaning ear of corn.

Charles A. Schweighauser
is professor of Astronomy-Physics and director,
Sangamon State University Observatories.